INTEGRATED LAKE-WATERSHED ACIDIFICATION

INTEGRATED
LAKE-WATERSHED ACIDIFICATION

Reprinted from Water, Air, and Soil Pollution, Vol. 26 No. 4 (1985)

D. Reidel Publishing Company / Dordrecht / Boston

Library of Congress Cataloging in Publication Data

CIP-data appear on separate card.

ISBN-13:978-94-010-8926-5   e-ISBN-13:978-94-009-5498-4
DOI:10.1007/978-94-009-5498-4

Published by D. Reidel Publishing Company,
P.O. Box 17, 3300 AA Dordrecht, Holland.

Sold and distributed in the U.S.A. and Canada
by Kluwer Academic Publishers,
190 Old Derby Street, Hingham, MA 02043, U.S.A.

In all other countries, sold and distributed
by Kluwer Academic Publishers Group,
P.O. Box 322, 3300 AH Dordrecht, Holland.

TABLE OF CONTENTS

INTEGRATED LAKE-WATERSHED ACIDIFICATION

INTEGRATED LAKE-WATERSHED ACIDIFICATION STUDY: SUMMARY

ROBERT A. GOLDSTEIN

Electric Power Research Institute, Environmental Assessment Department, P.O. Box 10412, Palo Alto, CA 94303, U.S.A.

CARL W. CHEN

Systech Engineering, Inc., 3744 Mt. Diablo Boulevard, Suite 101, Lafayette, CA 94549, U.S.A.

and

STEVEN A. GHERINI

Tetra Tech, Inc., 3746 Mt. Diablo Boulevard, Suite 300, Lafayette, CA 94549, U.S.A.

(Received November 1, 1984; revised May 14, 1985)

Abstract. An integrated, interdisciplinary, intensive study of three forested watersheds in the Adirondack Park region of New York State was started in 1977 to quantify the relationship between the deposition of atmospheric acids and surface water acidity. A general mechanistic theory of lake-watershed acidification that takes into account the production and consumption of acidity by watershed processes, as well as atmospheric inputs of acidity, was developed. This theory is formulated as a mathematical simulation model.

Model and data analyses establish the importance of using an integrated ecosystem perspective to assess the vulnerability of surface waters to acidification by acidic deposition. The acid-base status of surface waters is determined by the interaction of many factors: soil, hydrologic, vegetation, geologic, climatic, atmospheric. The absolute and relative contribution of any given factor can be highly variable, both geographically and temporally; hence, lake sensitivity to changes in the quality and quantity of atmospheric deposition is also highly variable.

1. Introduction

The Integrated Lake-Watershed Acidification Study (ILWAS) was started in 1977 in the Adirondack Park region of New York State in response to questions regarding the role that acidic depostion plays in determining the acidity of surface waters. On a theoretical basis, it is certainly feasible that the acidity of lake water can change in response to changes in the chemical composition of atmospheric deposition. An extreme example would be a lake in a monolithic granite basin with no soil or vegetation. Such a system would be similar to a glass beaker receiving direct precipitation. The pH of its water would directly reflect the acidity of the deposition.

From the perspective of the policy analyst, however, the interest is not in this simple case but in real systems with the added complications of vegetation, soil, etc. A major assessment question is not 'Can or have lakes been acidified by acidic deposition?', but 'How much acidification can or has occurred, and is it reversible?' By 'how much' is meant how many lakes have changed from one given pH value to another. A half unit decrease of pH in a pH 7.5 lake would probably be biologically insignificant, while a half unit decrease in a pH 5.5 lake could have major biological consequences.

The purpose of this set of papers is to summarize the major findings and results of ILWAS. Although the publication of these manuscripts marks the official completion of the study, given the richness and size of the collected data set (over 600,000 measurements), and the robustness of the mathematical model of lake-watershed adification developed by the study, meaningful analysis of both will probably continue for many years.

2. Study Objective

The objective of ILWAS was to develop a tool that could be used to assess the amount of acidification that can or has occurred as a result of atmospheric deposition; that is, to develop a quantitive relationship between the acidity of deposition and that of surface waters. This was accomplished by developing a general mechanistic theory of lake-watershed acidification that takes into account the production and consumption of acidity by watershed processes, as well as atmospheric inputs of acidity. The original conceptualization of this theory was described by Chen et al. (1979) and Goldstein et al. (1980). The theory is formulated as a mathematical model that simulates in an integrative manner the multiple processes controlling lake acidity (Gherini et al., 1985).

3. Study Design

An overview of the study design is given by Goldstein et al. (1984). The fundamental concept on which the design of ILWAS was based is that a lake's vulnerability to acidification by atmospheric deposition can only be understood in context of the acid-base chemistry and hydrology of its entire catchment. Based on this concept, the approach taken was to indentify the major flowpaths followed by precipitation before reaching a lake and the major processes that alter the chemical characteristics of water as it moves along these flowpaths (Figure 1).

The use of small watertight catchments as experimental units for ecosystem analysis can be traced back to the early hydrological studies of the U.S. Forest Service (Gaskin et al., 1983). Significant advances in the approach resulted from the Hubbard Brook Study (Likens et al., 1977) and the International Biological Program (IBP), (Reichle, 1981). Hubbard Brook focused on the simultaneous analysis of inputs and outputs of major ion species and water, while IBP emphasized cycling processes within watersheds. ILWAS, for four consecictive years, simultaneously investigated both internal processes and material inputs and outputs for three lake-watersheds.

The lake-watershed system was divided into a network of compartments: atmosphere, canopy, snowpack, vegetation, soil horizons, bogs, stream segments and lakes. Field investigations and laboratory experiments were designed to determine the processes controlling the movement of water through the compartments and the chemical composition of the water within the compartments. A mechanistic approach, in contrast to an input-output analysis, was taken so that the results of the study could be applied to sites other than those of the study. A mathematical simulation model was developed

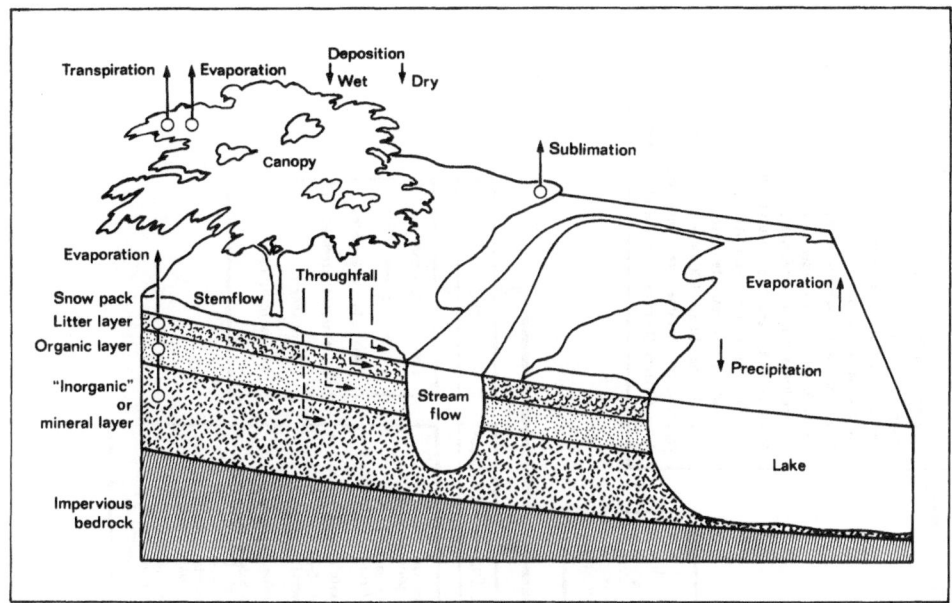

Fig. 1. Pathways of water tributary to a lake.

that served as a framework for integrating individual processes. The processes and chemical constituents simulated by the model are described by Gherini *et al.* (1985).

Over a period of four years (1977–1981), the quality and quantity of atmospheric deposition and water moving through the compartments of each of the three study watersheds were regularly measured. Data were also collected to characterize the biogeochemical properties of the watersheds. The modeling and data collection programs were executed concurrently to allow feedback between them and modification of both during the project period. The model was applied as an integrative tool to test consistency among data subsets, and between the model's conceptual basis and the data.

It is important to emphasize the long-term aspect of ILWAS. At the time of its inception, there were many individuals who claimed that because of policy considerations (i.e., the needs of policy makers) the technical questions addressed by ILWAS needed to be answered within a time frame of 6 mo.

However, it was recognized by those involved with the development of ILWAS, that such a short-term approach would be fruitless. Neither a 6 mo nor a 1 yr program could produce a technically sound and adequate data base and theoretical framework for developing a general quantitative relationship between acidic deposition and lake acidity. Hence technical considerations were factored into determining the length of the study. ILWAS is unique in terms of the length of time for which such intensive data collection has been conducted simultaneously for so many biogeochemical aspects of three watersheds, During the course of ILWAS, numerous short-term studies of surface water acidification have been conducted by others. None of these studies has defined

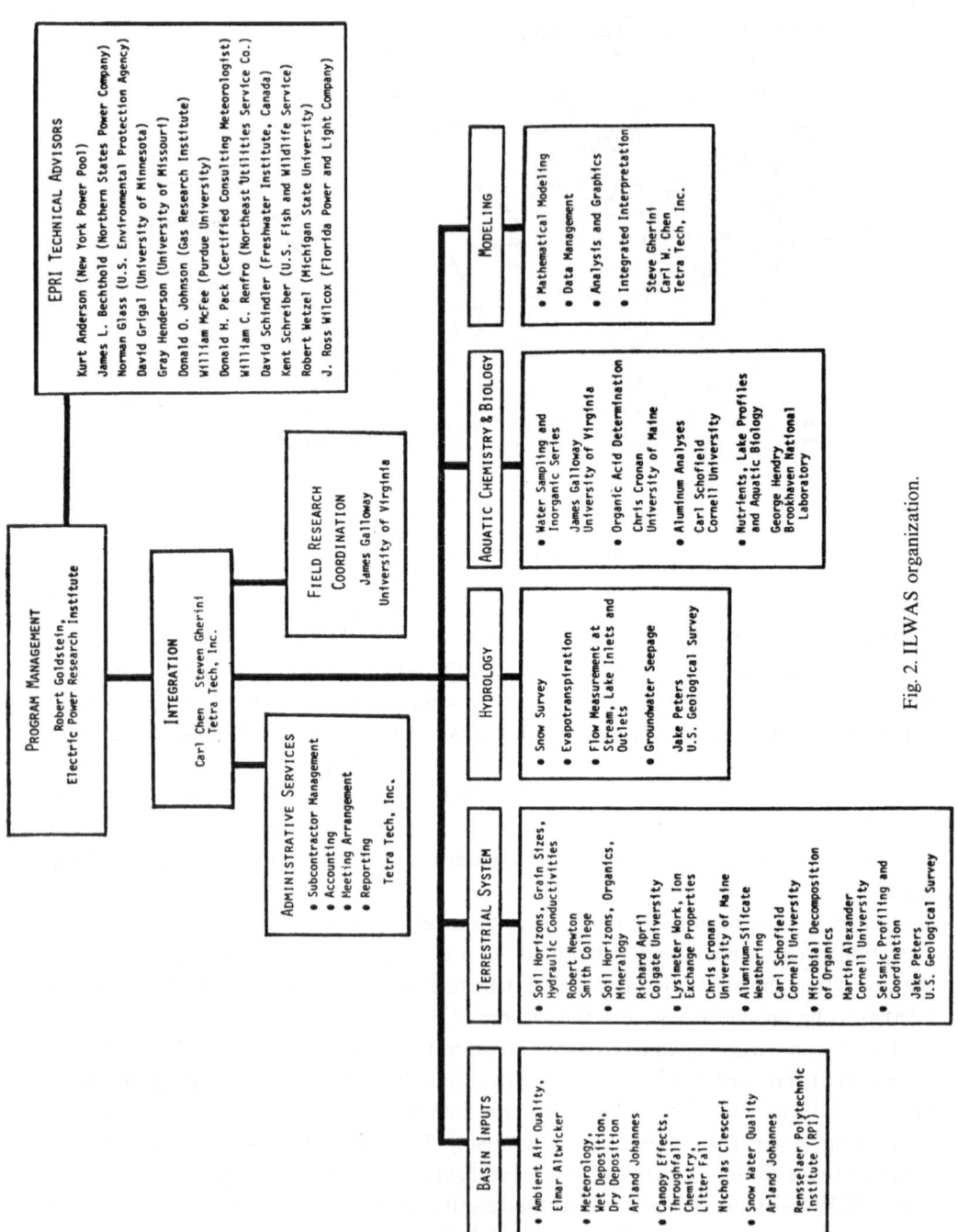

Fig. 2. ILWAS organization.

the general relationship between acidic deposition and lake acidity, nor resolved the technical questions addressed by ILWAS. The policy matters that supposedly had to be addressed within 6 mo are still waiting to be resolved. The results of ILWAS can play a major role in their resolution.

The scope of ILWAS spanned many disciplines: meteorology, plant ecology, soil science, geology, hydrology, limnology and aqueous chemistry. As a result, the ILWAS study team was interdisciplinary in nature. Its core was made up of 14 principal investigators from 10 institutions (Figure 2). In addition, there was an independent interdisciplinary group of advisors who periodically reviewed the study's progress. The adminstrative organization of the study was designed to reinforce the emphasis on technical integration (Figure 2).

4. Study Sites

The ILWAS field sites consisted of three forested watersheds, located within 30 km of each other in the Adirondack Park region of New York State (Figure 3). Although each watershed received about the same amount of precipitation of nearly identical quality (Johannes *et al.*, 1985), each had a lake with a different pH level and different pH dynamics (Figure 4). Of the three lakes, Woods was considered acidic, with a typical outlet pH between 4.5 and 5.0, and Panther neutral, with a typical outlet pH near 7. Sagamore Lake had a much larger watershed (Table I), with more spatially hetero-geneous biogeochemical characteristics, that results in more variable year-round pH dynamics, with an outlet pH that was typically between that of Woods and Panther.

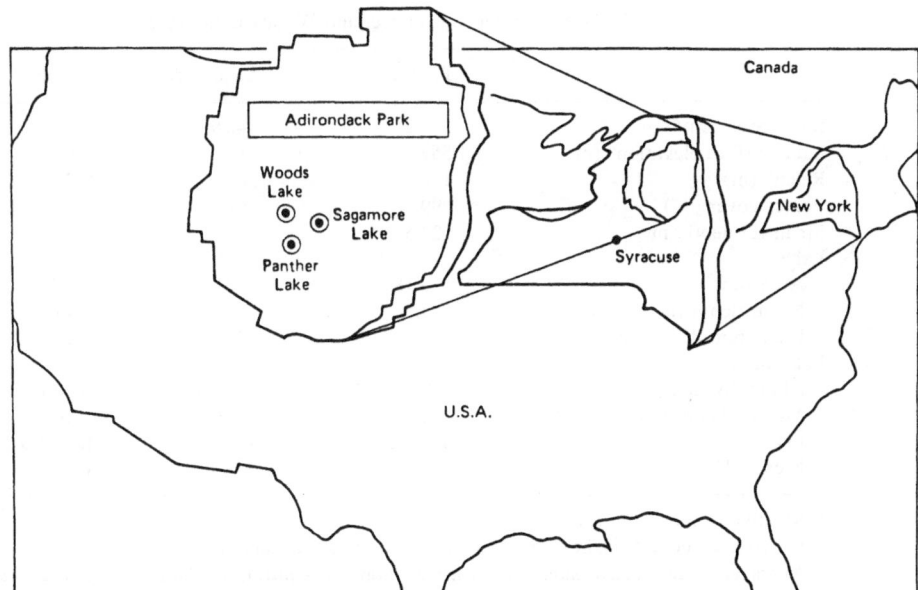

Fig. 3. Location of ILWAS field study areas.

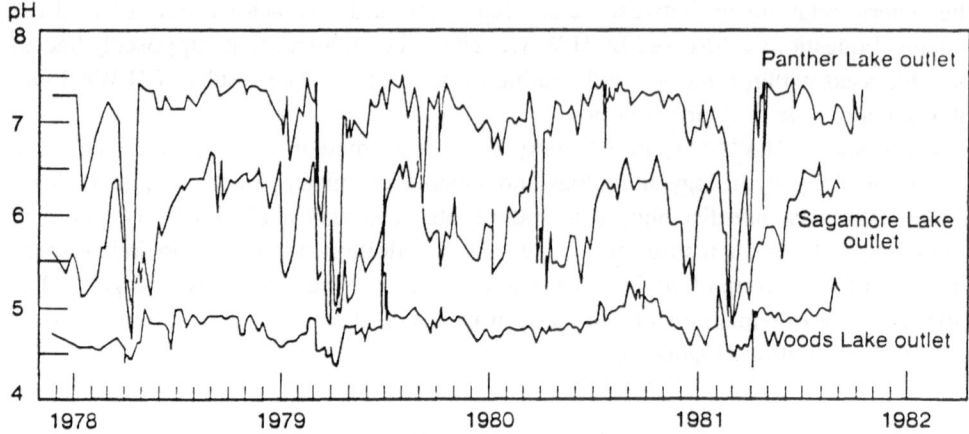

Fig. 4. Outlet air-equilibrated pH values for Woods, Sagamore, and Panther lakes. The depressions in pH which occurred in the spring (most pronounced in Panther) were associated with the spring snowmelt period and decreases in base cation concentrations and increases in the nitrate in the lake surface waters.

For detailed discussions of the chemistry of the lakes, the relationship between lake and outlet chemistry, and the depression in outlet pH that occurred at the time of spring snowmelt (Figure 4), see the paper by Schofield *et al.* (1985), and also Galloway *et al.* (1984), and Goldstein *et al.* (1984).

TABLE I

Characteristics of Panther, Sagamore, and Woods Lake watersheds

	Panther	Sagamore	Woods
Basin area (km^2)	1.2	48.9	2.1
Lake surface elevation[a] (m)	557	580	606
Relief[b] (m)	170	561	122
Forest cover (%)	99	91	96
Mean till depth, m	24.5	–	2.3
Lake			
Area (km^2)	0.18	0.66	0.26
Mean depth (m)	3.5	8.8	4.0
Maximum depth (m)	7	23	12
Lake outlet			
Alkalinity (μeq L^{-1})	– 35 to 240	– 30 to 80	– 60 to 30
Mean alkalinity (μeq L^{-1})	147	31	– 10
pH	4.5–7.2	4.7–6.5	4.4 – 5.9
Mean pH[c]	6.2	5.6	4.7

[a] Relative to mean sea level.
[b] Difference between highest and lowest elevation in the watershed.
[c] Averaged after conversion to hydrogen ion concentration. Note these are field pH values, not air-equilibrated values as shown in Figure 4. In general, air-equilibrated pH values are higher due to the loss of volatile weak acids such as aqueous CO_2.

General study site characteristics are listed in Table I. Details regarding atmospheric deposition, vegetation, soils, hydrology, geology, and water chemistry can be found in the papers of Johannes *et al.* (1985), Peters and Murdoch (1985), April and Newton (1985), Cronan (1985), and Schofield *et al.* (1985) – all in this report.

At the beginning of the study, most attention was focused on the Sagamore watershed, based on the assumption that Sagamore Lake was in transition from a neutral to an acidic state and that the rate of acidification was sufficiently great that an increase in acidity could be observed within the time span of the study. This assumption proved to be incorrect. No interannual trend in acid-base status of any of the lakes was observed during the course of the study (Figure 4). As the study evolved, Woods and Panther watersheds became the sites of major interest, because given the radically different acid-base status of their lakes and the small and relatively homogeneous nature or their catchments, the comparison of their properties provided the greatest insight into defining the roles of different hydrologic and biogeochemical processes in determining lake acidity. Hence the papers which follow emphasize analyses that compare and contrast the Woods and Panther systems.

5. Results

There exists a widespread misconception concerning surface water acidification. Many people believe that, in general, a single factor (e.g., atmospheric deposition or organic acids of natural origin) determines the acid-base status of lakes. Furthermore, these people anticipate that a study such as ILWAS will identify this factor, at least for the Adirondacks. In fact, what ILWAS clearly demonstrates (in the following papers) is that acid-base status is determined by the interaction of many factors (e.g., soil, hydrologic, vegetation, geologic, climatic, atmospheric) and that the relative role of any one factor can be highly variable geographically (even with a region as small as that defined by the ILWAS sites) and temporally. Temporal variation is illustrated by the drop in Panther Lake outlet pH when the relative amount of water which flows through the organic soil horizon to the lake increases during snowmelt (Goldstein *et al.*, 1984; Peters and Murdoch, 1985).

Variability in factors determining the acid-base status of surface waters leads to differences in the ability of lake-watersheds to neutralize acid deposition. Although each of the ILWAS lake-watershed systems was a net sink for acidity (i.e., annual atmospheric input of acidity exceeds annual lake outflow), the amount of neutralization that occurred was highly variable among watersheds (Schofield *et al.*, 1985; Figure 16, in Gherini *et al.*, 1985) and with time (e.g., snowmelt period versus periods of low flow). The major sources of alkalinity within the study sites appeared to be ion exchange within the soils and mineral weathering (Newton and April, 1985; Figure 16, in Gherini *et al.*, 1985).

The rate at which hydrologic basins supply alkalinity to the through-flowing water establishes their vulnerability to acidification. (The alkalinity supply rate depends on many factors, including the acidity of the applied deposition; for example, weathering rates can increase in response to increased H^+ concentration.) All of the results of ILWAS confirm the major hypothesis on which the design of the study was based; that is, a lake's vulnerability to acidification by atmospheric deposition can only be understood in an ecosystem context that incorporates the biogeochemistry and hydrology of its entire catchment. Climatic and lake-watershed characteristics interact in multiple ways to determine vulnerability. Within a given geographical region such as the Adirondacks, there exists a great variety of lake-watershed systems, and hence considerable variability with respect to vulnerability. A lake such as Panther is relatively intensitive to changes in the acidity of atmospheric deposition, while Woods Lake is relatively sensitive (Figure 17, Gherini et al., 1985). It should be noted that the sensitivity of Panther varies seasonally, with sensitvity being greatest at the time of spring snowmelt.

It is important to recognize that during spring snowmelt only the surface waters of the lakes were acidified (Figure 4 in Schofield et al., 1985). The increase in acidity was related to a dilution of base cations and an increased concentration of nitrate (Schofield et al., 1985). The relative contribution of soil versus atmospheric sources to the nitrate pulse is currently being studied. Aqueous Al, Fe, and Mn also occurred at higher concentrations during the snowmelt period.

The analyses of soil and lake water chemistry support the concept of using alkalintity (acid neutralizing capacity) as the key variable for characterizing the acid-base status of surface waters (for a discussion of the underlying theory, see Gherini et al., 1985). Furthermore, to understand in general the acid status and dynamics of surface waters, it is necessary to understand the behavior of all major anions and cations (aqueous chemistry is discussed in Schofield et al., 1985; Gherini et al., 1985; Cronan, 1985; Galloway et al., 1984). Simple empirical relations based on the concentrations of a few ions (e.g. Henriksen, 1979, 1980) do not appear to be generally applicable (Kramer and Tessier, 1982; Church and Galloway, 1984). For instance, sulfate concentration by itself cannot in general be used as a surrogate for acidity. Woods and Panther lakes had nearly identical average annual sulfate concentrations, about 125 μeq L^{-1} (Schofield et al., 1985), but very different average annual pH and pH dynamics. Also during the snowmelt transient in pH, sulfate concentration remained relatively constant but nitrate, Al and base cation concentrations changed dramatically (Figures 7, 9, 10, and 11, Schofield et al., 1985).

Hydrologic and chemical analyses lead to the general conclusion that the relative routing of water through a watershed is a major determinant of lake water alkalinity and vulnerability to acidification by atmospheric deposition. A lake may be located in a watershed underlain with readily weatherable alkaline rock (e.g. limestone), but if the water flowing into the lake is isolated from the rock by impermeable soils then the rock cannot supply alkalinty to the lake. The difference in acid status between Woods and Panther lakes and the snowmelt acid pulse can both be explained by flowpath analysis (Peters and Murdoch, 1985). Panther Lake was more alkaline than Woods because water flowing through the deep soils made a greater relative contribution to its inflow. (Figure 6, April and Newton, 1985; Goldstein et al., 1984). During snowmelt, the high application rate of water to the catchments caused the lateral ground water flow to 'back up' and the relative routing of water in the watersheds to change so that the percentage of water flowing into the lakes through the shallow, more acidic organic horizon increased (Goldstein *et al.*, 1984). The major sources of base cations and hence alkalinity in the ILWAS watersheds, cation exchange and mineral weathering, were in the deeper inorganic soil. Of course, it should be kept in mind that flowpath is an integrating concept (not a single factor) and is a function of lake watershed characteristics (e.g., soil hydraulic conductivity, land slope) and environmental conditions (e.g., precipitation rate, air temperature).

For the ILWAS study sites, the relative contribution of the deep soil flowpath is directly correlated to soil depth. Thus soil depth appears to be a good single key indicator for differentiating the acid-base status of the three lakes (Figure 13, Gherini *et al.*, 1985; Goldstein *et al.*, 1984). However, it is essential to recognize that soil depth would not necessarily be a good single key indicator in more general situations. We have observed, in watersheds with highly permeable deep soils, seepage lakes (lakes without surface inlets and outlets) that are isolated from the groundwater by relatively impermeable sediments. Hence it is possible to have, in watersheds with deep soils, lakes whose water inflow is almost entirely precipitation directly onto the lake. Such a lake could be highly sensitive to changes in the acidity of atmospheric deposition even though the underlying soils contained a large volume of water with high alkalinity.

Given the variability in possible responses of surface water to identical levels of acidic deposition, assessment on a regional basis will ultimately require the derivation of frequency distributions of response (Figure 5). Lake-watersheds within a given region

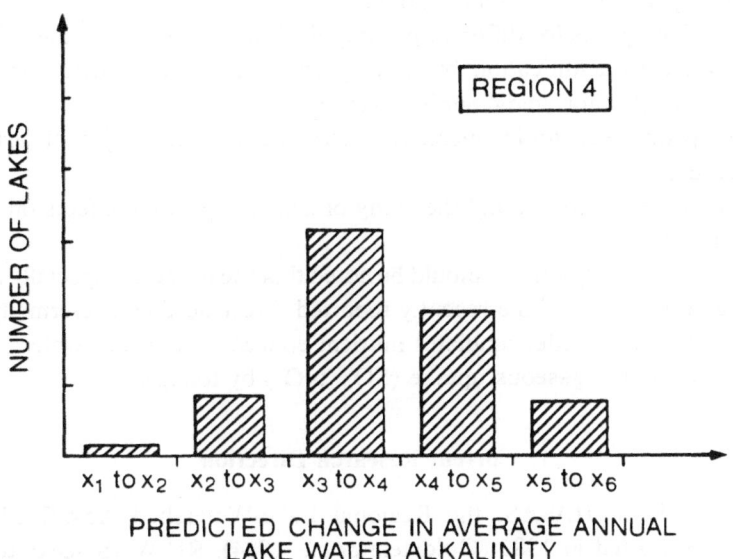

Fig. 5. Hypothetical reponse of average annual alkalinity of lakes within a region to halving of total sulfur deposition.

will have to be divided into different response classes based on biogeochemical and hydrologic characteristics. Data will have to be collected for a representative sample of the various lake-watershed types within the region. This problem is being addressed by the follow-on study to ILWAS that will be described in the last section of this paper.

6 Application of Model

To evaluate the effectiveness of various pollution control and acidification mitigation strategies, it is necessary to characterize quantitatively how nonacidic lakes would react over time to continuing acidic depositon at current levels and how all lakes would respond over time to changed levels of acidic deposition and various neutralization schemes. ILWAS was a research study and did not produce environmental assessments per se. It did, however, develop the ILWAS model which can be used in making such assessments.

Analysis of ILWAS data has lead to increased understanding of the processes that control lake-watershed acidification. However, the data cannot necessarily be used to assess how surface waters other than the study lakes would respond to possible future levels of acidic deposition. The ILWAS model, though, can be applied to data from other watersheds in the Adirondacks and other regions to make such assessments. Hence, the model has potentially the most widespread utility of the ILWAS results. Widespread use of the model should lead to increased robustness and better definition of the uncertainties inherent in its use.

Major potential applications of the model include:

(1) Preparation of data analyses on which technically sound policy decisions can be based regarding surface water acidification.

(2) Design of programs for different geographical regions to assess the vulnerability of surface waters to acidification by atmospheric acids and to determine effective management strategies for acidic surface waters.

(3) Use as a pedagogical tool to increase understanding of the subject of surface water acidification, and

(4) Use as a research tool to aid the study of acidic deposition effects on terrestrial and aquatic biota.

With respect to the last point, it should be noted that the model is especially applicable to researching the causes of the recently reported forest decline in Germany and the eastern U.S., since the model simulates nutrient concentrations and water availability in the rooting zone and gaseous uptake (SO_2, NO_x) by foliage.

7. Current Research Direction

A follow-on study to ILWAS, the Regional Lake-Watershed Acidification Study (RILWAS) was started in 1982 (Goldstein et al., 1984). RILWAS seeks to test and increase the robustness of the ILWAS model by applying it to areas outside of the Adirondacks and to develop and test a methodology to assess the vulnerability of an entire region (in contrast to a few individual lakes) to acidification by atmospheric acids. The first objective is being addressed by studying watersheds in northwestern Wisconsin, western North Carolina, and the Sierra Nevada Moutains. The second objective is being addressed in the Adirondacks, where 20 study watersheds have been selected to represent the variety of environmental situations that exist in the region.

References

April, R. and Newton, R. M.: 1985, *Water, Air, and Soil Poll.* **26**, 373 (this volume).

Church, M. R. and Galloway, J. N.: 1984, *Water, Air, and Soil Pollution* **22**, 111.

Chen, C. W., Gherini, S. A., and Goldstein, R. A.: 1979, in M. J. Wood (ed.), *Ecological Effects of Acid Precipitation* (EA-79-6-LD), Electic Power Research Institute, Palo Alto, California.

Cronan, C. S.: 1985, *Water, Air and Soil Poll.* **26**, 355 (this volume).

Galloway, J. N., Altwicker, E. R., Church, M. R., Cosby, B. J., Davis, A. O., Hendrey, G., Johannes, A. H., Nordstrom, K. D., Peters, N. E., Schofield, C. L., and Tokos, J.: 1984, *The Integrated Lake-Watershed Acidification Study*, Volume 3: Lake Chemistry Program. Electric Power Research Institue, EA-3221, Volume 3, RP1109-5.

Gaskin, J. W., Douglass, J. E., and Swank, W. T.: 1983, 'Annotated Bibliography of Publications on Watershed Management and Ecological Studies at Coweeta Hydrologic Laboratory, 1934, 1984', SE-30, Southeastern Forest Experiment Station, Asheville, North Carolina.

Gherini, S. A., Chen, C. W., Mok, L., Goldstein, R. A., Hudson, R. J. M., and Davis, G. F.: 1985, *Water, Air and Soil Poll.* **26**, 425 (this volume).

Goldstein, R. A., Chen, C. W., Gherini, S. A., and Dean, J. D.: 1980, in D. Drabløs and A. Tollan (eds.), *Ecological Impact of Acid Precipitation*, SNSF Project, Oslo, Norway.

Goldstein, R. A., Gherini, S. A., Chen, C. W., Mok, L., and Hudson R. J. M.: 1984, *Phil. Trans. R. Soc., Lond. B* **305**, 409.

Henriksen, A.: 1979, *Nature* **278**, 542.

Henriksen, A.: 1980, in D. Drabløs and A. Tollan (eds.), *Ecological Impact of Acid Precipitation*, SNSF Project, Oslo, Norway.

Johannes, A. H., Altwicker, E. R., and Clesceri N. L.: 1985, *Water, Air, and Soil Poll.* **26**, 339 (this volume).

Kramer, J., and Tessier, A.: 1982, *Environ Sci. Technol.* **16**, 606.

Likens, G. E., Boerman, F. H., Pierce, R. S., Eaton, J. S., and Johnson, N. M.: 1977, *Biogeochemistry of a Forested Ecosystem*, Springer-Verlag, New York, 146 pp.

Peters, N. E. and Murdoch, P. S.: 1985, *Water, Air, and Soil Poll.* **26**, 387 (this volume).

Reichle, D. E. (ed.): 1981, *Dynamic Proporties of Forest Ecosystems*, Cambridge University Press, Cambridge, 683 pp.

Schofield, C. L., Galloway, J. N., and Hendry, G. R.: 1985, *Water, Air, and Soil Poll.* **26**, 403 (this volume).

THE INTEGRATED LAKE-WATERSHED ACIDIFICATION STUDY: ATMOSPHERIC INPUTS

ARLAND H. JOHANNES

School of Chemical Engineering, Oklahoma State University, Stillwater, OK 74078, U.S.A.

ELMAR R. ALTWICKER, and NICHOLAS L. CLESCERI

Department of Chemical Engineering and Environmental Engineering, Rensselaer Polytechnic Institute, Troy, NY 12181, U.S.A.

(Received November 1, 1984; revised June 5, 1985)

Abstract. Atmospheric inputs to Woods, Panther, and Sagamore Lake-Watersheds in the Adirondack Mountains of New York State were measured on a daily basis from March 1978 through December 1981. Precipitation quality was nearly identical at all sites on monthly and yearly bases; ion loadings to each watershed were principally controlled by the amount of precipitation. No yearly trend was evident for any ion concentration in wet deposition. Annual precipitation quantities showed little deviation from long-term averages for this region. Throughfall measured under various species of trees showed enrichment in most base cations and acid anions. Deciduous trees were found to increase the pH of incident precipitation, while coniferous canopies tended to decrease pH.

1. Introduction

The Integrated Lake-Watershed Acidification Study (ILWAS) was initiated in late 1977 to quantify the relationship between acidic deposition and the acidity of surface waters. Three lake-watersheds in the Adirondack Mountains of New York were chosen for intensive study based on surveys which showed significant differences in pH among the lakes. (Typical lake water pH values: Woods 4.7, Sagamore 5 to 6, Panther 7.) Since the three lake-watersheds, Panther, Woods, and Sagamore are located within 30 km of each other, a hypothesis of the study was that the three lake-watersheds receive similar amounts of precipitation with nearly identical chemical composition.

The main objective of this element of the ILWAS project was to quantify wet and dry deposition, throughfall chemistry, and ambient air quality, and to determine whether the three lake-watersheds did in fact receive similar atmospheric inputs.

Estimates of both wet and dry deposition were also needed to make overall watershed elemental balances. Although measurement techniques for wet depositon are relatively straightforward, no routine method for monitoring dry deposition is currently accepted by all scientists. Measurements of bulk deposition, throughfall and stemflow, and dryfall, using a dry bucket, were made throughout the study to estimate dry deposition and investigate the effect of a forested canopy on incident precipitation.

2. Methods

From March 1978 through December 1981, atmospheric inputs were measured for the Woods, Panther and Sagamore lake-watersheds. Atmospheric inputs in the form of rain, snow, and dry deposition were collected using Aerochem Metrics wet/dry collectors. These collectors consisted of metal frame and housing for two collection buckets – a 'wetfall bucket' and 'dryfall bucket'. A sliding roof, activated by a moisture sensor, kept the wetfall bucket covered except during precipitation, when the roof would move to cover the dryfall bucket.

The collection buckets are constructed of white polyethylene and are 28.6 cm in diameter and 24.8 cm deep with a slight taper from top to bottom. The collectors were mounted on a wooden platform 2 m high. Next to each collector a standard weighing bucket rain gage was installed to obtain accurate precipitation quantity data. Two collectors were placed in each watershed, in open areas away from trees (Figure 1). In addition, a seventh collector was located near the field laboratory in Big Moose, New York. Wind speed, wind direction, and ambient temperature were monitored in Woods and Panther watersheds.

Rain, snow, throughfall and bulk samples were collected on a daily basis. Dryfall samples were collected weekly.

Wet samples were transported in covered collection buckets to the field laboratory where measurements were made of sample volume, field pH, and conductivity. The remainder of the sample was then filtered through 0.4 µm nuclepore filters and stored in clear polyethylene bottles. Samples were refrigerated at 3 °C without preservatives pending chemical analysis. Dryfall samples were extracted with 250 mL of distilled-deionized water and processed in the same manner.

Samples were analyzed for pH, conductivity, SO_4^{2-}, NO_3^-, Cl^-, NH_4^+, Ca^{2+}, Mg^{2+}, Na^+, and K^+. The cations (except NH_4^+) were measured using a Perkin Elmer Model 403 Atomic Absorption Spectrophotometer. Anions and NH_4^+ were analyzed using Technicon Auto Analyzers. Specific details of the sampling network design, chemical analyses, and quality control procedures are reported elsewhere (Johannes et al., 1980, 1981, 1985).

3. Results

3.1. PRECIPITATION QUANTITY

From the end of November until April, snow and mixed precipitation predominate in the Adirondacks. However, rain occurred during the January-February mid-winter thaws. Ninety-five percent of the precipitation events, including snow storms, ranged from 0.02 to 6.0 cm of water equivalent depth. Precipitation occurred about every third or fourth day. Over the study period there was no apparent seasonal pattern of dry or wet months.

Annual precipitation measured in each basin was similar over the investigation, with

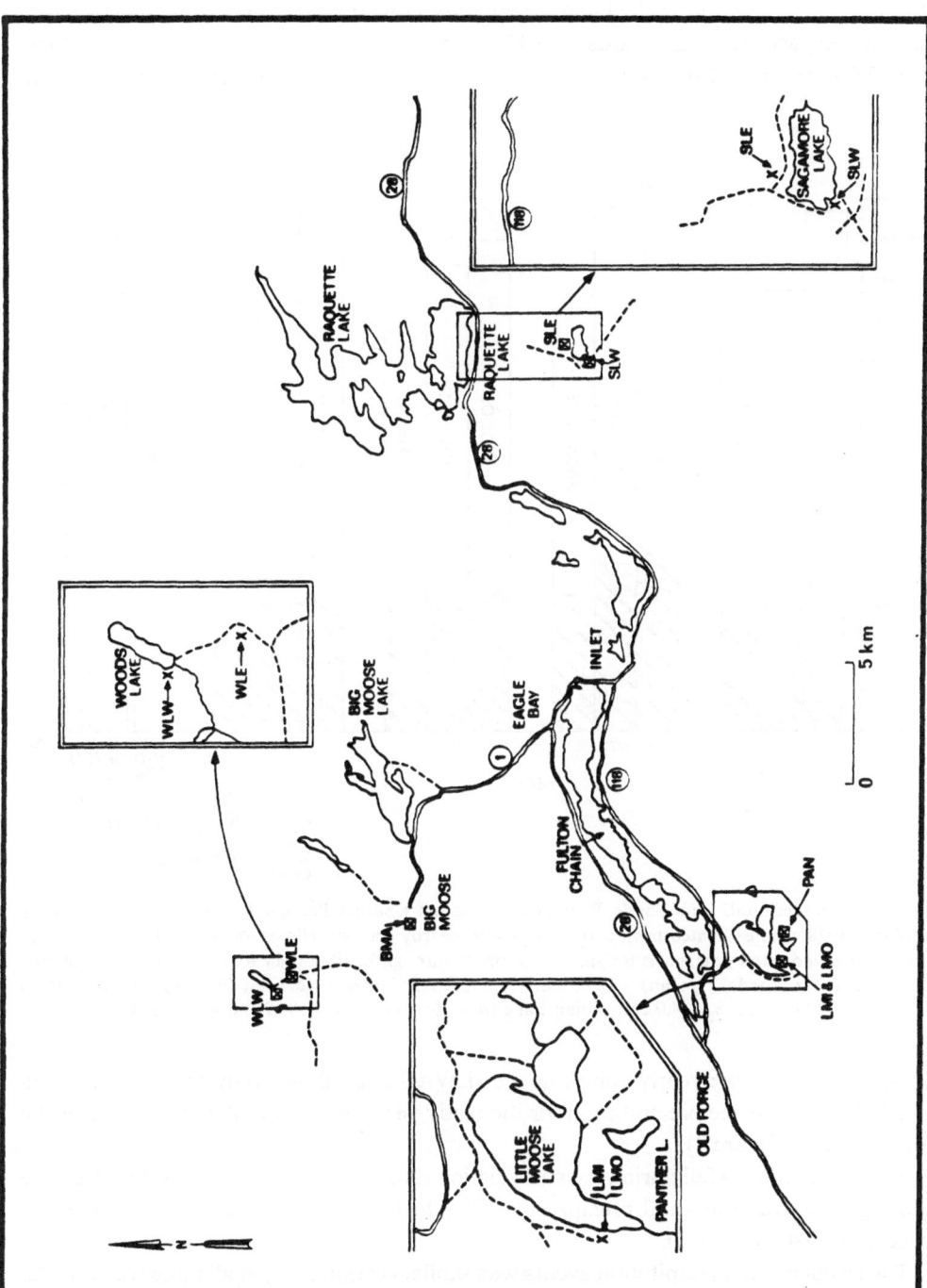

Fig. 1. Sampling station locations. Station identification key: BMA – Big Moose; LMI – Little Moose Inner; LMO – Little Moose Outer; PAN – Panther; SLE – Sagamore Lake East; SLW – Sagamore Lake West; WLE – Woods Lake East; WLW – Woods Lake West.

the possible exception of a somewhat drier 1980 (Figure 2). Ten-year means and standard deviations were calculated for the three closest NOAA precipitation collection stations: Big Moose (near Woods) $\bar{x}$ = 126.8 cm, σ = 16,6 cm; Old Forge (near Panther) $\bar{x}$ = 12.06 cm, σ = 14.0; Indian Lake (near Sagamore) $\bar{x}$ = 106.2 cm, σ = 15.8 cm.

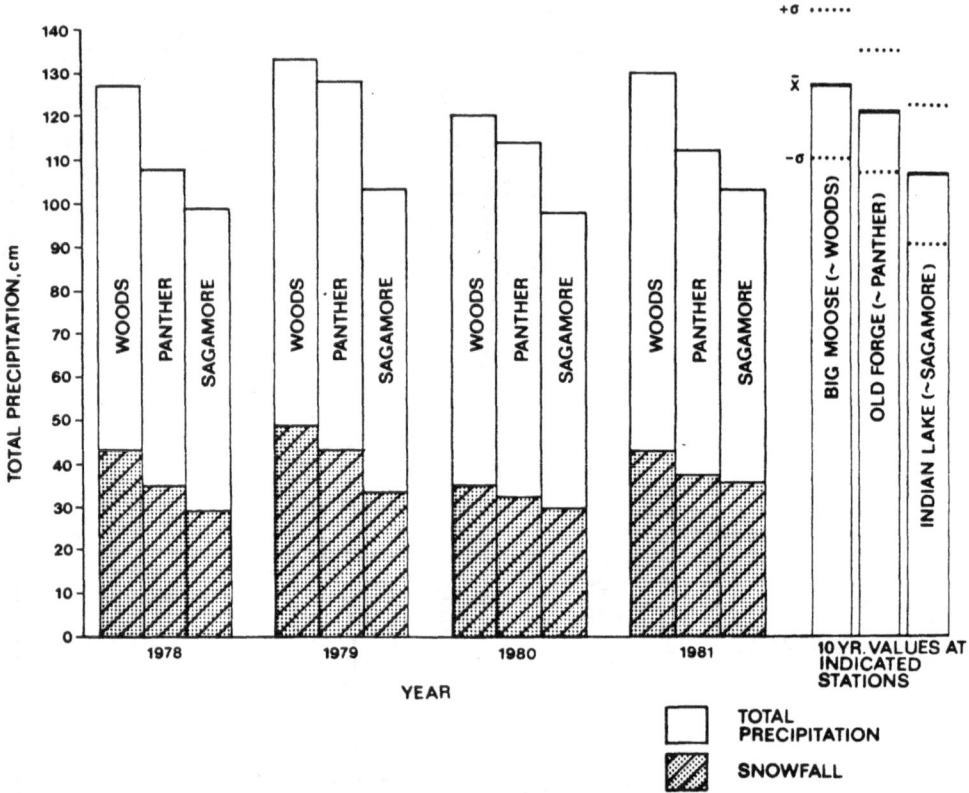

Fig. 2. Annual precipitation in Woods, Panther, and Sagamore basins. Precipitation amounts from January to March 1978 were estimated from correlations with nearby stations. (Basis for totals: January through December of indicated year; basis for snow: November through April of indicated year.) NOAA stations: Big Moose (near Woods) $\bar{x}$ (mean) = 126.8 cm, σ (standard deviation) = 16.6 cm; Old Forge (near Panther) $\bar{x}$ = 120.6 cm, σ = 14.0 cm; Indian Lake (near Sagamore) $\bar{x}$ = 106.2 cm, σ = 15.8 cm.)

Comparison with the yearly values at the ILWAS sites shows only minor differences and indicates that the precipitation for the study years was 'normal' for this area of the Adirondacks (Figure 2).

Total annual snowfall during the study period (November through April) followed the same general trend as total precipitation. The highest snowfall occurred in 1979, the lowest in 1980 (Figure 2).

The frequency of precipitation events was similar (within 5 %) in all three watersheds. During the 45-mo study there were 418 events at Woods, 416 events at Panther, and 397 events at Sagamore. Yet, Sagamore received 21% less precipitation than Woods

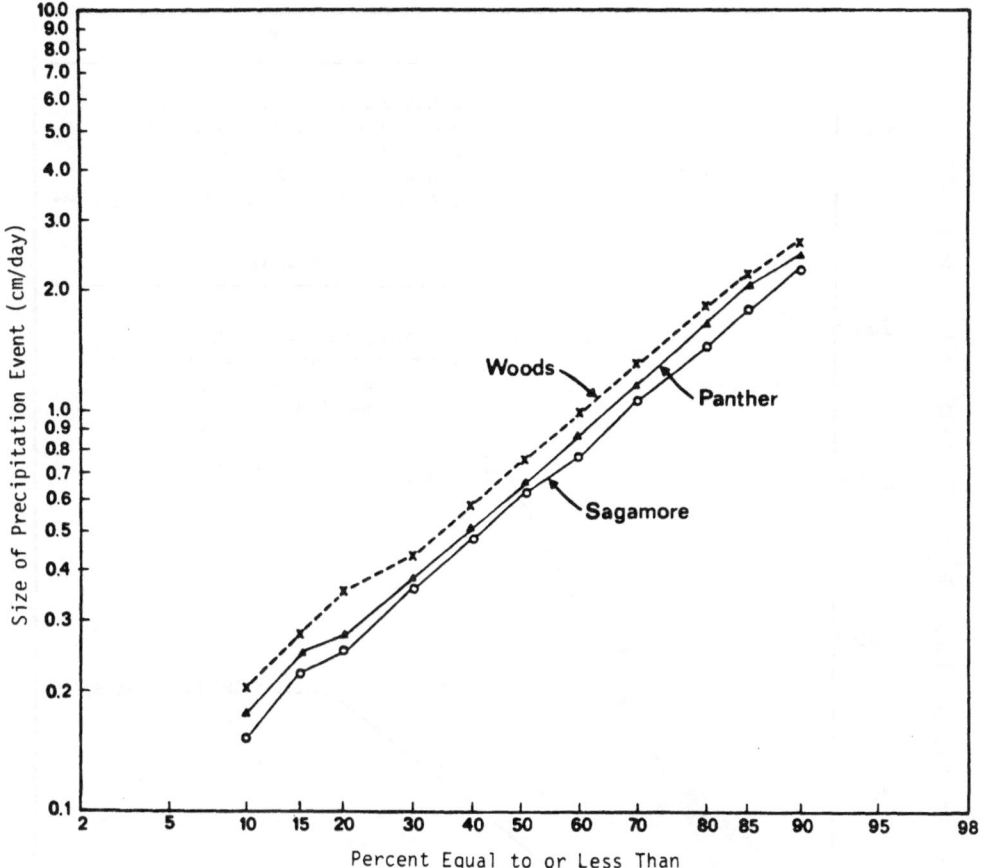

Fig. 3. Distribution of the sizes of precipitation events in Woods, Panther, and Sagamore basins.

and 14% less than Panther. A cumulative frequency distribution for precipitation event size is shown in Figure 3. Event size was lognormally distributed for all basins. Over the 45-mo study period, Woods received 480 cm of precipitation, Panther 440 cm, and Sagamore 380 cm. The lower total precipitation at Sagamore was due mainly to lower precipitation *volume* per event.

3.2. ION CONENTRATION-WETFALL

The sum of base cations was plotted versus the sum of acid anions for regions which have high (>4.5) and low (<4.5) precipitation pH (Figure 4). Precipitation with excess strong acid anions relative to base cations is acidic and falls to the right of the 1:1 line. The ratio of the sum of base cations to the sum of acid anions in ILWAS precipitation is close to 1:2.5 (Figure 4). The study region can, therefore, be classified as an area with relatively acidic precipitation.

Event concentrations for all ionic species versus time are exemplified by the pH of precipitation measured at Woods Lake (Figure 5). No long-term trend in concentration

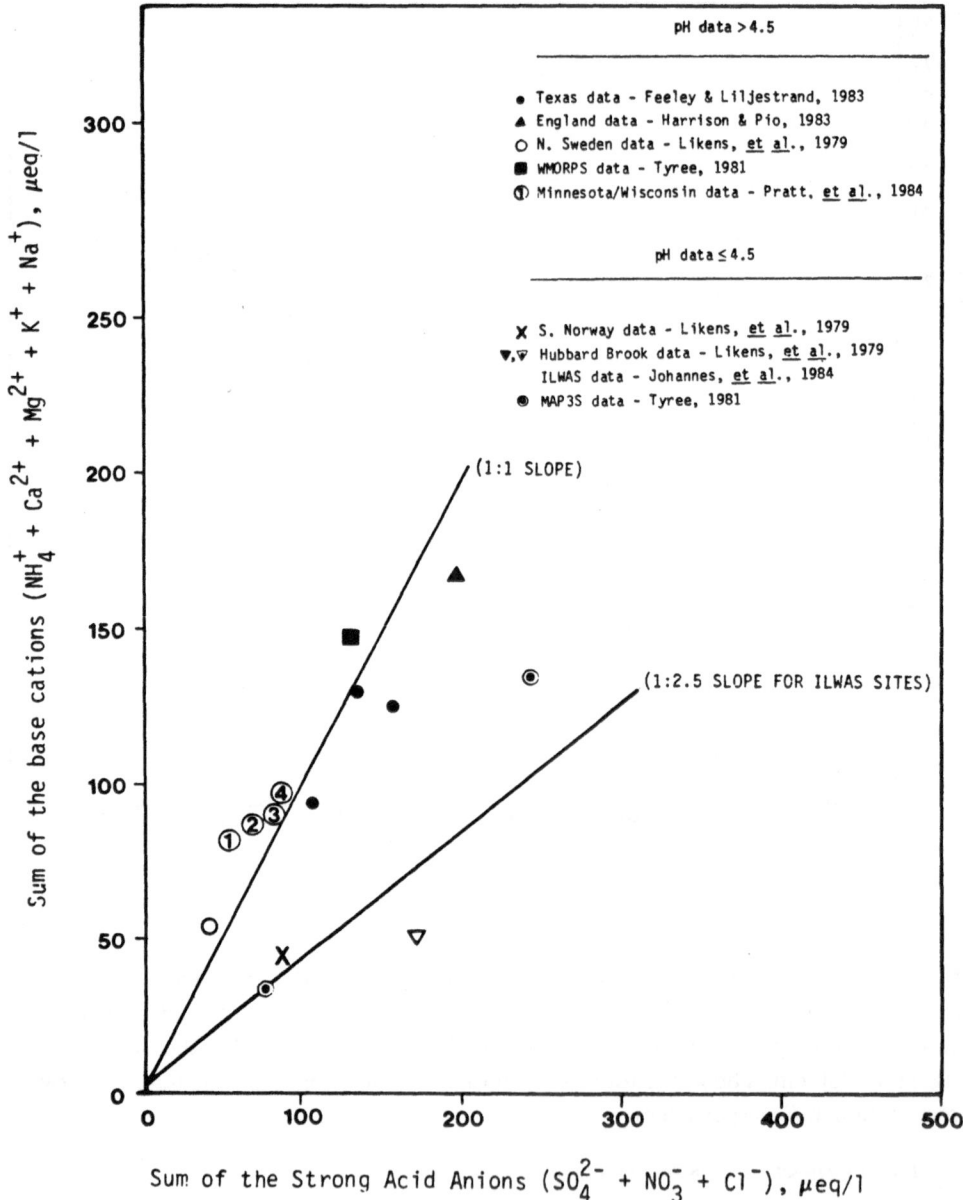

Fig. 4. Sum of base cations versus sum of acid anions in precipitation. Precipitation with excess strong acid anions relative to base cations is acidic.

was evident in any ion concentration over the 45-mo study period. However, order of magnitude differences in solute concentrations did occur between successive events and even between collection sites for individual events.

Volume-weighted ion concentrations were calculated for each site and ion on a

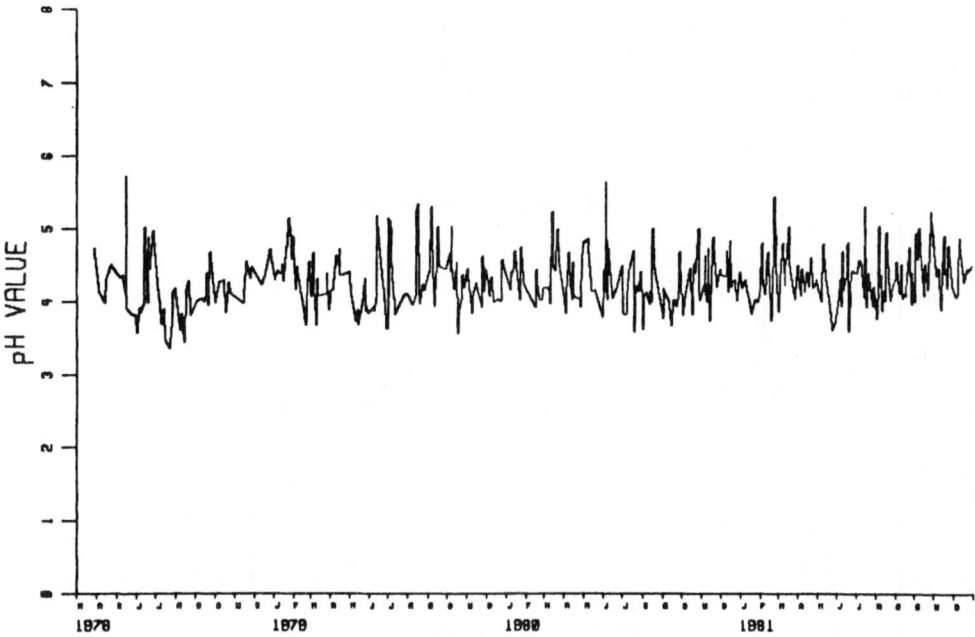

Fig. 5. pH values of precipitation events at Woods Lake (WLW).

monthly, seasonal, and yearly basis. Volume-weighted ion concentrations for each watershed, on an annual basis, were nearly identical.

Mean monthly volume-weighted ion concentrations were also calculated for all ions from weighted concentrations from all sites (Johannes *et al.*, 1985). The standard deviations were divided by the means of the monthly values to calculate coefficients of variation (CV). These values are compared to similarly calculated CV values reported by Galloway and Likens (1978) and by Granat (1976) in Table I. Values for the ILWAS sites appear to be reasonable given the station spacings used.

TABLE I

Coefficients of variation for wet deposition measurements

| | Galloway and Likens (1978) | Granat (1976) | ILWAS | |
			May 1978 – Aug. 1979	Sep. 1979– Dec. 1981
Diameter of Region, km	5	100	1–30	1–30
H^+	0.09	–	0.03–0.20	0.05–0.13
SO_4^{2-}	0.09	0.29	0.03–0.18	0.04–0.21
NO_3^-	0.09	0.32	0.06–0.20	0.04–0.18
NH_4^+	0.13	0.87	0.06–0.37	0.07–0.25

Mean monthly volume-weighted ion concentrations for the five major ions in precipitation (H^+, SO_4^{2-}, NO_3^-, NH_4^+, and Ca^{2+}) are compared in Figure 6. Sulfate, the major acid anion, showed a yearly summer peak in concentration. Nitrate was less

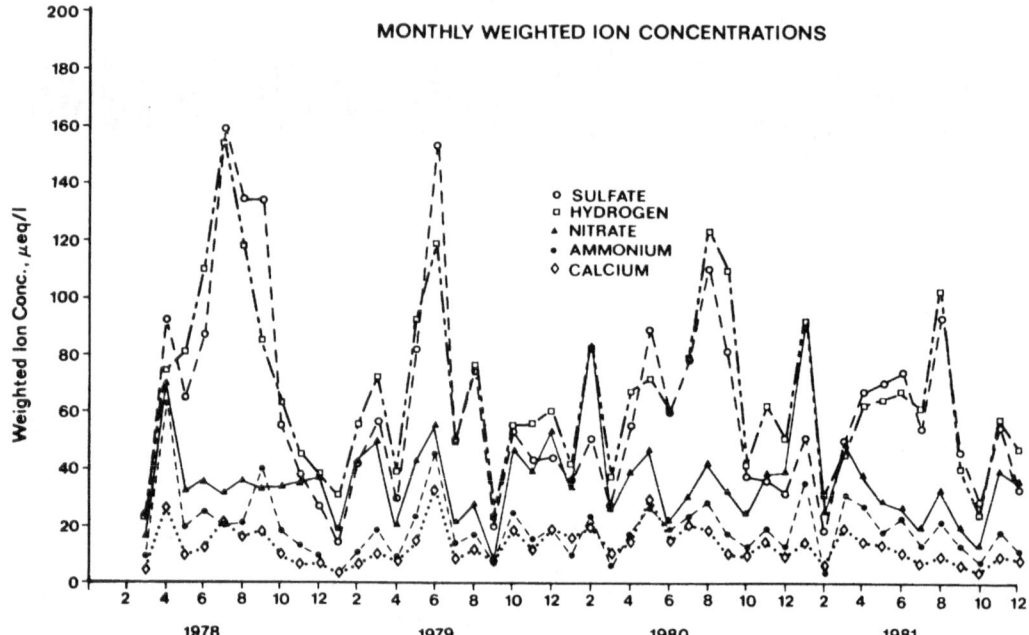

Fig. 6. Mean volume-weighted monthly ion concentrations in precipitation as measured at Woods, Sagamore, and Panther lakes.

variable and reached its maximum during the winter. Hydrogen-ion concentrations were highly correlated with sulfate ($r^2 = 0.83$, $n = 45$) and showed similar summer peaks. These relationships were similar on an event basis at all sites. Figure 7 shows the relationship between precipitation pH and strong acid anion concentrations. The minor ions, Cl^-, Mg^{2+}, Na^+, and K^+, generally occurred at concentrations less than 10 μeq L^{-1} and showed no apparent yearly trends.

Since precipitation quality was nearly identical at all sites on a monthly or yearly basis, while precipitation quantity had a greater variability, the wet loadings (eq ha^{-1}yr^{-1}) to each watershed were principally controlled by the *quantity* of precipitation received.

3.3. TOTAL DEPOSITION

Total deposition is defined as the sum of wet and dry deposition, where dry deposition includes both gaseous and particulate matter. However, the methods used to quantify dry deposition produce values with large uncertainties. Although several methods can be used to estimate dry deposition, the focus here will be on dry bucket measurements.

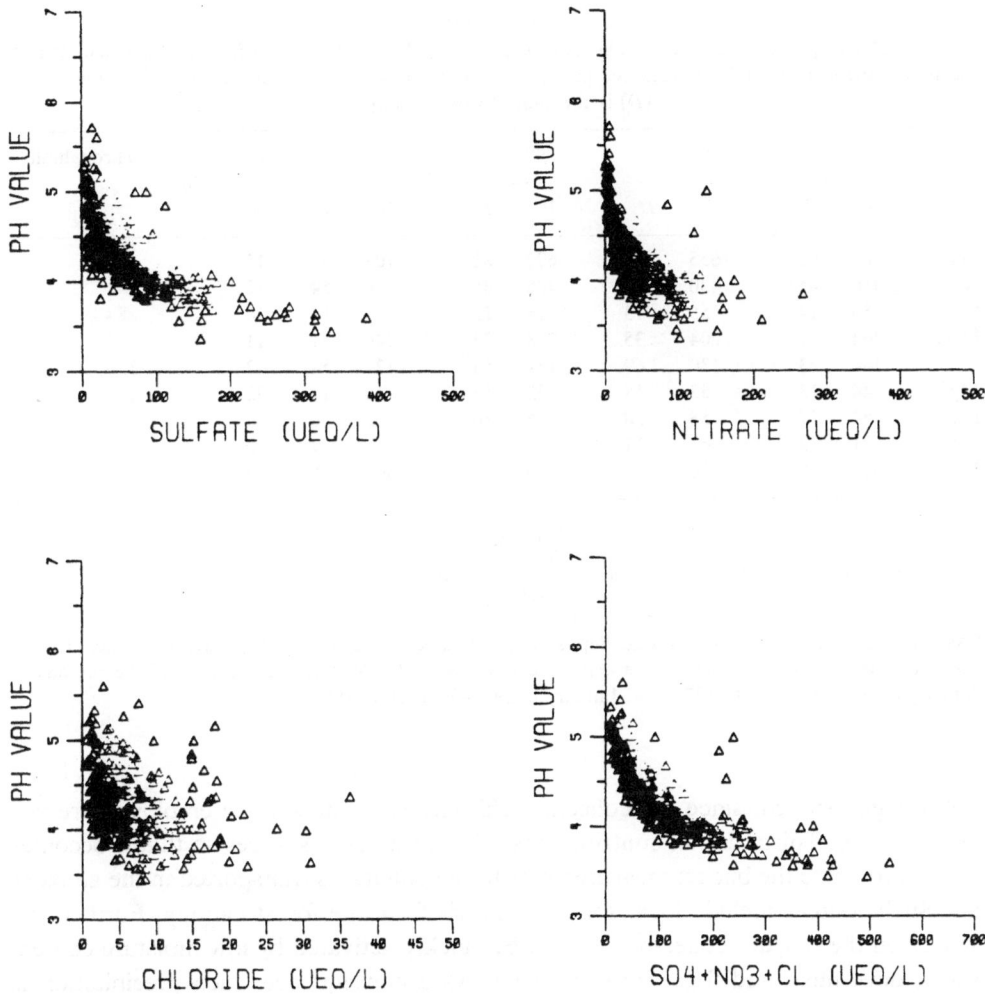

Fig. 7. Relationship between pH and strong acid anion concentrations in precipitation at Woods Lake.

Dry deposition as measured using dryfall buckets was a significant fraction of total deposition (Table II). The weekly variability between watersheds was considerably greater for dry deposition than for wet. However, on a yearly basis, the dry deposition was very similar in each watershed.

The key question that is not answered by such comparisons is the representativeness of the values obtained from the dry-bucket collectors. First, the values are only time-averaged values and not 'event' values. Second, the interpretation of the values is

TABLE II

Wet and Dry Deposition at Woods Lake Values are eq ha^{-1} yr^{-1}. Wet deposition (W) is calculated as the sum of the products of the measured precipitation quantities and ion concentrations. Dry deposition (D) is the sum of weekly samples.

	1978		1979		1980		1981		D[a] as a % of W	Throughfall[b]
	W	D	W	D	W	D	W	D		
SO_4^{2-}	880	92	655	118	673	82	708	89	13	1080
NO_3^-	460	45	430	61	408	36	380	54	12	–
Cl^-	53	14	55	13	56	12	57	14	24	84
NH_4^+	264	21	204	35	273	24	244	31	11	–
Ca^{2+}	144	63	130	69	164	53	121	54	43	330
Mg^{2+}	44	13	53	18	52	15	36	14	32	165
Na^+	51	17	58	18	55	16	62	16	30	101
K^+	21	12	43	11	23	10	18	9	40	600
H^+	914	49	670	97	826	36	608	45	8	–

[a]Dry deposition as a % of wet deposition: $\dfrac{\sum\limits_{4 \text{ yr}} \text{dry dep}}{\sum\limits_{4 \text{ yr}} \text{wet dep}} \times 100$

[b]As upper bounds on the actual total deposition, simulated throughfall values have been included for selected ions. Simulations were made using the ILWAS model. Values are averages of the periods of January 8, 1978–January 8, 1979, and January 8, 1979–January 8, 1980.

not straightforward, since the collection efficiencies of the surface employed are not really known. Third, dryfall contributions arise from many sources so that it becomes difficult to relate the bucket measurements to the pollutants transported in the ambient air. Such sources include local soils and debris from the forest canopy. Further, the sensor on the wet/dry collector may not be quickly activated by low moisture content snow, dew, mist, and fog events; if at all. As a consequence, such precipitation is collected (at least in part) in the dry bucket and becomes part of the dryfall. It is the authors' opinion that dew, mist, and fog events should be considered part of dry deposition since most wet-only collectors do not capture this flux.

A reasonable interpretation of dryfall data would be to view dry bucket loadings as minimum values for dry deposition. Gravitational settling, diffusion, and impaction are probably the major removal mechanisms, and these are known not to be the most efficient for small particle size ranges. The dry bucket offers a relatively stagnant environment in a sheltered watershed, and the approach velocity with which air masses move through forests is virtually absent inside a bucket. The forest canopy, because of its large surface area, and the lake surface, because of the higher efficiency of a wet collecting surface, are most likely more efficient collectors than a dry bucket.

Values for throughfall fluxes for Woods Lake have been added to Table II as upper limits on total deposition. Throughfall includes both wet and dry deposition but may also include foliar exudation or leaching of some elements. Therefore, loadings based upon throughfall should be used with care when estimating net basin inputs (see Throughfall, Section 3.4).

Ambient air quality measurements were made to check dryfall measurements. Measurements were made every 5 to 7 days and were limited to total suspended particulates (TSP) and particulate sulfate and nitrate. Sulfate and nitrate were also segregated into fine and coarse fractions using a dichotomous sampler.

Probability distributions for TSP, sulfate, and nitrate concentrations at Woods and Sagamore Lakes are shown in Table III. Most of the sulfate ($>80\%$) appeared in the fine (<2.5 µm) fraction; particulate nitrate was equally distributed between coarse and fine fractions. Seasonal patterns were evident in TSP and sulfate levels, with summer peaks and winter minimums. Little seasonal dependence was observed in nitrate concentrations.

A limited number of particulate Ca concentrations were measured and averaged 0.57 µg m^{-3}. Also, shown in Table III are distributions for gaseous SO_2 and HNO_3. These compounds were measured on an episodic basis in 1980 and 1981 using integrating

TABLE III

Distributions of ambient air concentrations at Woods (BMA) and Sagamore lake stations

Item and station	Ambient air concentration µg m^{-3} at the percentile[b]		
	20%	50%	80%
Total suspended particles			
Woods Lake (BMA)	8	15	26
Sagamore Lake (SLE)	6	13	24
Sagamore Lake (SLW)	6	15	33
Composite[c]	7	14	30
Sulfate			
Woods Lake (BMA)	1.4	2.9	6.0
Sagamore Lake (SLE)	1.4	2.9	6.0
Sagamore Lake (SLW)	1.2	2.3	5.0
Composite	1.3	2.6	5.5
Nitrate			
Woods Lake (BMA)	0.07	0.17	0.43
Sagamore Lake (SLE)	0.06	0.20	0.70
Sagamore Lake (SLW)	0.05	0.17	0.45
Composite	0.06	0.17	0.45
Gaseous SO_2[a]	0.80	1.30	3.10
Gaseous HNO_3[a]	0.24	0.40	0.52

[a] Gaseous concentrations measured episodically in 1980 and 1981.
[b] Non-exceedance frequencies for HIVOL sampler data.
[c] Composite determined from data at BMA, SLE, and SLW stations.

collection methods (Johannes *et al.*, 1984). Maximum observed values were 3.7 µg m^{-3} for SO_2 and 5.9 µg m^{-3} for HNO_3.

Using these ambient air concentrations and deposition velocities, comparisons were made with dry bucket measurements. Good agreement was found during summer months but winter comparison in general were poor due to contamination of the dry bucket with snow. A detailed comparison on several different time scales is available (Johannes *et al.*, 1983).

3.4. THROUGHFALL

Precipitation over a forested watershed is altered by a brief but significant interaction with plant surfaces, which collect airborne gases and particles and release exudates. These processes can result in major changes in the quality of incident precipitation as it passes through the canopy and becomes throughfall.

Throughfall was collected under canopies which had been shown by a detailed survey (Cronan, 1985) to be representative of the different watersheds: yellow birch, American beech, sugar and red maple, red spruce, and balsam fir. Throughfall samples were collected on an event basis in the Woods Lake watershed. Collector design and analytical methods have been reported elsewhere (Vasudevan and Clesceri, 1983).

During the period November-April, when deciduous trees are without leaves, the pH-values of throughfall under deciduous species were similar to those of incident

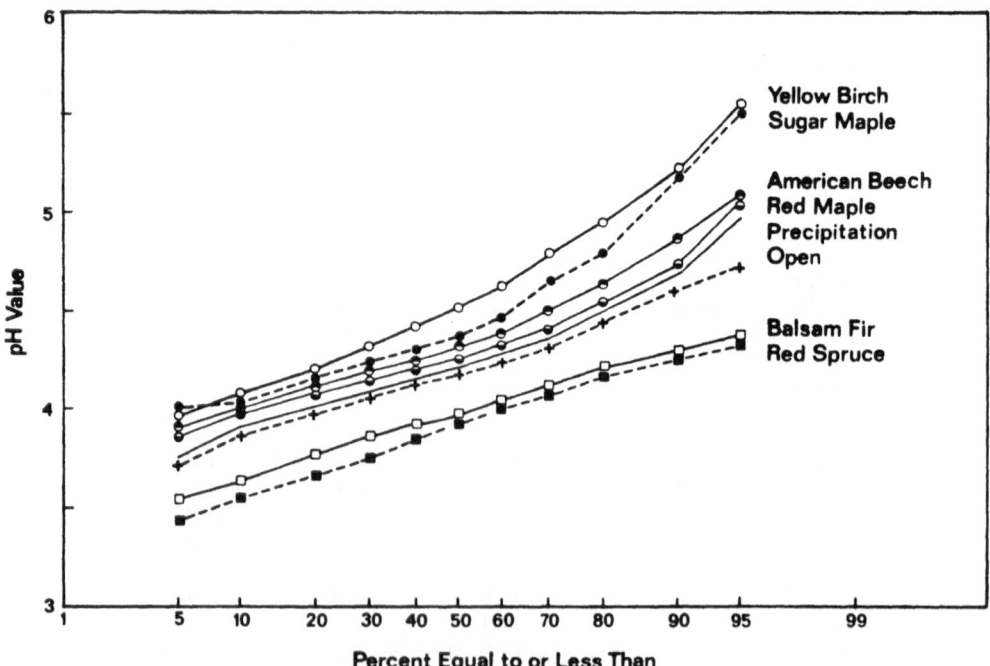

Fig. 8. Distribution of pH values in throughfall and precipitation samples. Open samples were taken from bulk collectors, placed in clearings.

precipitation (pH = 3.5 to 4.5). During the summer, the throughfall was less acidic (pH = 4.0 to 6.5) under deciduous trees and showed greater enrichment of base cations over acid anions. By contrast, the throughfall collected under coniferous canopies during summer was generally more acidic than that of the incident precipitation and showed a greater enrichment of acid anions over base cations. Enrichment, however, was found in all ions except hydrogen (as noted above) and ammonia.

Probability distributions of throughfall pH were nearly lognormal and showed differences among tree species relative to wet only or bulk precipitation (Figure 8).

Potassium concentration was greatly enriched in throughfall and showed a definite seasonal pattern. In the summer, concentrations reached 150 to 200 μeq L^{-1}, depending on the canopy type. In incident precipitation, the concentration was very low ($< 10\ \mu$eq L^{-1}); dry deposition of K was also very small. The marked increase in throughfall K can be attributed to foliar exudation.

Throughfall sulfate concentrations were higher than those observed in precipitation, but the increase depended on tree species. In the summer, sulfate concentrations ranged from 300 to 400 μeq L^{-1} under deciduous canopies while coniferous canopies produced concentrations up to 600 μeq L^{-1}. Sulfate concentrations in winter throughfall under coniferous canopy were similar to those found during the summer. Enrichment of sulfate

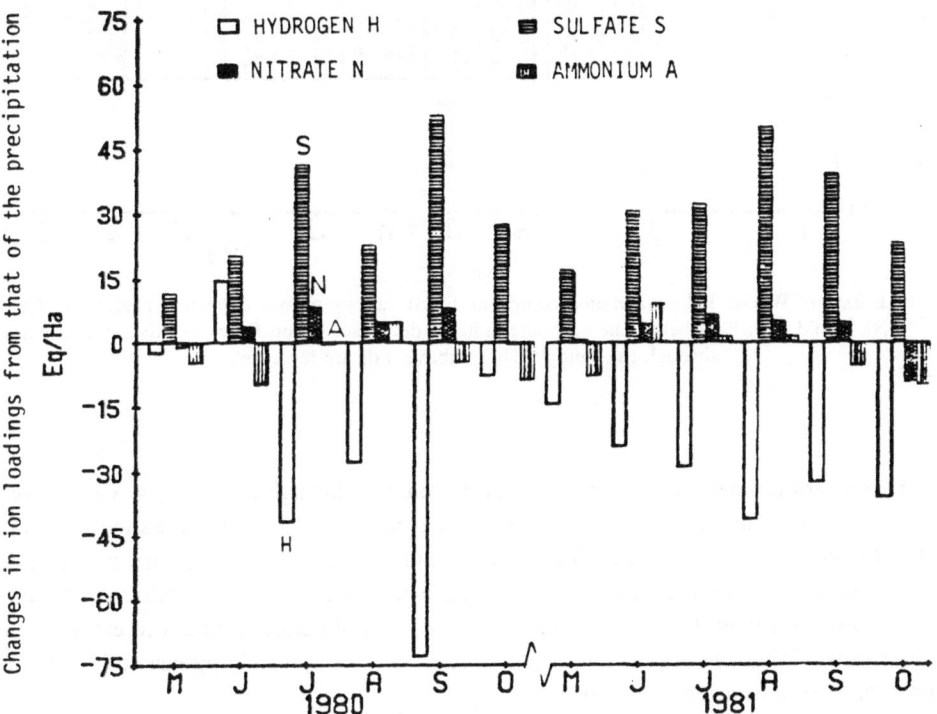

Fig. 9. Effect of Woods Lake watershed composite forest canopy on wet deposition loadings of H^+ – ion, SO_4^{2-}, NO_3^-, and NH_4^+. Bars below the zero line indicate decreases in ion fluxes as precipitation passed through the canopy. Those above indicate increases.

in throughfall is mainly due to dry deposition. Differences between canopies appear to result from leaf surface area and collection efficiency.

Figures 9 and 10 show the effect of the Woods Lake canopy (predominately deciduous) on wet depositon. Bars below the zero line indicate decreases in ion flux and

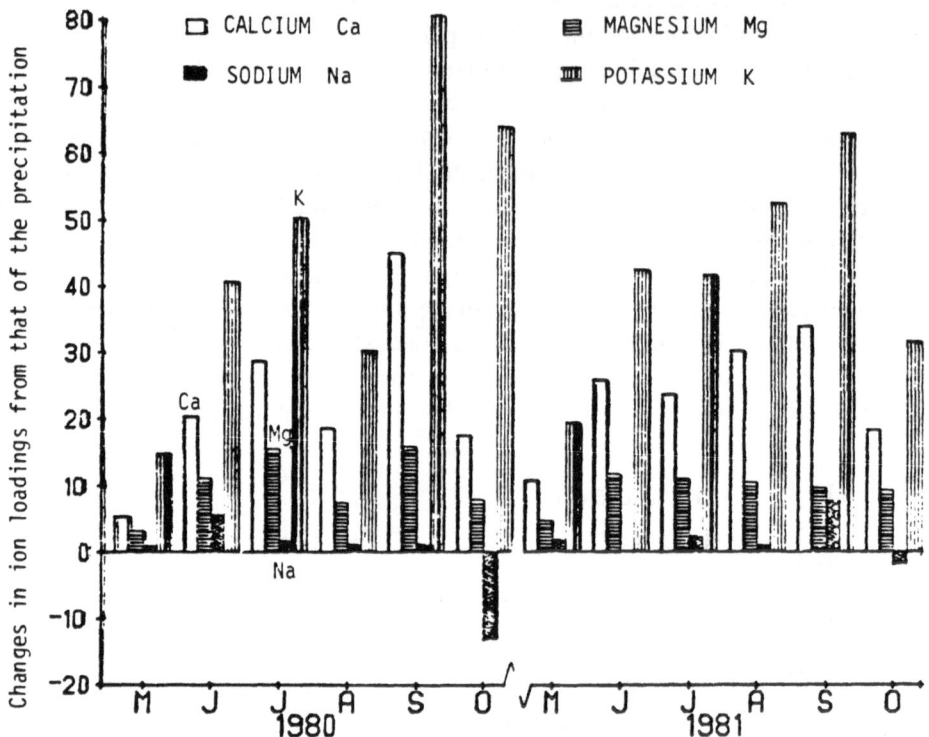

Fig. 10. Effects of Woods Lake watershed composite forest canopy on wet depostion loadings of Ca^{2+}, Mg^{2+}, Na^+, and K^+. Bars below the zero line indicate decreases in ion fluxes as precipitation passed through the canopy. Those above indicate increases.

those above indicate increases as precipitation passes through the canopy. Values were calculated using the percent coverage by each tree species in the watershed (Cronan, 1985). In general, passage of precipitation through this 'composite' canopy reduced the H^+ and NH_4^+ concentrations but increased the concentration for all other major ions.

Care must be taken, however, when using throughfall concentrations to estimate net basin inputs. Foliar exudation of elements recycled from the soil appears to be a major source of base cation enrichment in throughfall.

4. Conclusions

(1) During the study, annual precipitation quantities showed little deviation from long-term averages for this region.

(2) Differences between watersheds in total annual precipitation were mainly due to the size rather than the frequency of the events.

(3) Precipitation quality was nearly identical for the three watersheds on a monthly or yearly basis.

(4) Differences between watersheds in ion loadings (wet deposition) was principally controlled by the quantity of precipitation.

(5) Data support the hypothesis that the ion loading from wet depositon on an annual basis to all three basins were comparable. Wet deposition loadings to Woods and Panther were more similar and slightly higher than the loadings to Sagamore.

(6) No long-term trend was evident for any ion concentration in wet deposition.

(7) Hydrogen-ion concentration was highly correlated with sulfate concentration.

(8) Dry deposition was a significant part of total deposition.

(9) Throughfall showed enrichment in most base cations and acid anions.

(10) Deciduous trees were found to increase the pH of incident precipitation.

(11) Coniferous trees were found to decrease the pH of incident precipitation.

References

Cronan, C. S.: 1984, *Water, Air, and Soil Poll.* **26**, 355 (this volume).

Feeley, J. A. and Liljestran, H. M.: 1983, *Atmos. Envir.*, **17**, 807.

Galloway, J. N. and Likens, G. E.: 1978, *The Collection of Tellus*, **30**, 71.

Granat, L.: 1976, *Principles in Network Design for Precipitation Chemistry Measurements*, Proc. Symp. on Atmospheric Contribution to the Chemistry of Lake Waters, J. Great Lakes Res., Volume 2(1), p. 42.

Harrison, R. M. and Pio, C. A.: 1983, *Atmos. Envir.* **17**, 2539.

Johannes, A. H. and Altwicker, E. R.: 1980, 'Atmospheric input into Three Adirondack Lake Watersheds', in *Ecological Impact of Acid Precipitation*, Proc. of an International Conference, Sandefjord, Norway, March 11–14, 1980; SNSF project, pp. 256.

Johannes, A. H., Altwicker, E. R. and Clesceri, N. L.: 1981, *Characterization of Acidic Precipitation in the Adirondack Region*, Electric Power Research Institute, Palo Alto, California, EA-1826.

Johannes, A. H. and Altwicker, E. R.: 1983, 'Relationships Between Dry Deposition as Measured Via Collection with a Dry Bucket versus Ambient Air Concentrations', in *Proceedings of the Fourth International Conference on Precipitation Scavenging*, Dry Deposition and Resuspension, Volume 2, p. 903, Elsevier.

Johannes, A. H. and Clesceri, N. L.: 1985, *The Integrated Lake-Watershed Acidification Study*, volume 4: Basin Input Analyses, Electric Power Research Institute, Palo Alto, California.

Likens, G. E., Wright, R. F., Galloway, J. N., and Butler, T. J.: 1979, *Scientific Am.* **241**, 43.

Pratt, G. C., Coscio, M. R., and Krupa, S. V.: 1984, *Atmos. Envir.* **18**, 173.

Vasudevan, C. and Clesceri, N. L.: 1983, 'Analysis of Throughfall Sampling Procedures', in *the Proceedings of the ILWAS Annual Review Conference*, Electric Power Research Institute, EA-2827, p. 31.

BIOGEOCHEMICAL INFLUENCE OF VEGETATION AND SOILS IN THE ILWAS WATERSHEDS

CHRISTOPHER S. CRONAN

Department of Botany and Plant Pathology, University of Maine at Orono, Orono, ME 04469 U.S.A.

(Received November 1984; revised July 2, 1985)

Abstract. The ILWAS catchments (Panther, Sagamore, and Woods) contain closely related variants of the northern hardwood-spruce-fir complex of the Adirondack Region. Dominant species in these watersheds are: American beech, sugar maple, red spruce, red maple, and yellow birch. On an areal basis, the watersheds contain 57 to 88% hardwood cover type and range in percent coniferous cover from 28% in Sagamore watershed to 5% in Woods catchment. Mean live basal area values range from 22 to 30 $m^2 ha^{-1}$ between catchments, while mean live stem densities range from 1400 to 1700 stems ha^{-1}. Weighted average leaf area indices for the watersheds range from a low of 5.2 in Woods to a high of 7.2 in Sagamore. The higher leaf area in Sagamore watershed may enhance the collection of dry deposition and may explain the higher SO_4^{2-} concentrations in surface waters within that system.

Soils in the ILWAS watersheds are predominantly Becket series Haplorthods, Fragiorthods, and Haplaquods. Soil chemical properties are almost identical in the acid lake catchment (Woods) and the circumneutral lake catchment (Panther). The soils are characterized by a low percent base saturation (generally < 8% B.S.), exhibit pH values ranging from 2.9 in the forest floor to 4.7 in the BC horizon, and contain relatively low concentrations of soluble and adsorbed SO_4^{2-}.

SO_4^{2-} is the dominant solution anion in upper soil layers within all three watersheds and in the lake waters of Woods and Sagamore basins. However, nitrate concentrations are also unusually high in soil solutions, suggesting that the systems may be N-saturated with respect to atmospheric inputs. Levels of organic acidity are elevated in surface horizon solutions and decline significantly with soil depth. In general, soil water chemistries in the Woods and Panther catchments are almost indistinguishable. Thus, the differences in lake pH and alkalinity cannot be explained on the basis of inter-watershed contrasts in soil chemistry and soil solution chemistry. Much of the explanation can be related to inter-watershed differences in hydrologic flow paths, coupled with distinct differences in solution chemistry between separate soil horizons and till strata.

1. Introduction

The Integrated Lake-Watershed Acidification Study (ILWAS) was initiated to develop a quantitative mechanistic understanding of the relationship between atmospheric deposition and surface water quality and acidity (Goldstein *et al.*, 1984). The study was conducted using three forested watersheds (Panther, Sagamore, and Woods) in the central Adirondack Park of New York State. Although these catchments receive similar loadings of acidic deposition (Johannes *et al.*, 1985), the three systems exhibit distinct differences in lake water pH (Schofield *et al.*, 1985) and alkalinity (Panther Lake = pH 6.2, Sagamore Lake = pH 5.6, and Woods Lake = pH 4.7). By comparing the biogeochemical behavior of these watersheds, the ILWAS investigators hoped to elucidate and to quantify the major ecosystem parameters controlling the fate of strong acids introduced from the atmosphere to lake-watershed systems.

The terrestrial portion of the ILWAS research program included an examination of the effects of forest vegetation, soil chemistry, and soil solution composition on surface water chemistry in the ILWAS systems. The vegetation studies were designed to contrast the quantitative patterns of tree community structure between the three Adirondack catchments. In the soil chemistry investigation, the objective was to quantify important soil chemical differences in the upper 1 m of surficial material in the three ILWAS watersheds. Finally, the soil solution studies were used to track the fate of atmospheric deposition in the ILWAS watersheds and were intended to evaluate the importance of soil processes in controlling lake water acidity/alkalinity patterns. Other aspects of the terrestrial research program are described in related papers by Peters and Murdoch (1985) and April and Newton (1985).

2. Study Sites

The ILWAS watersheds are located within 30 km of each other in the central Adirondack Park region of New York State (Goldstein et al., 1985). This heavily forested region of the Northeast is dominated by second growth hemlock – white pine – northern hardwood forest, with smaller amounts of spruce-fir forest dominating the high and low elevations, rocky ridgetops, and cool, damp water courses (Braun, 1950). Climate for the region is classified as Dfb by the Köppen-Geiger system, with a mean annual temperature of 5 °C and total annual precipitation averaging 100 to 120 cm. The watersheds are underlain by granitic bedrock and are mantled by variable depths of glacial till (April and Newton, 1983). Soils in the catchment are predominantly Becket Spodosols.

3. Methods

3.1. VEGETATION

Vegetation sampling and analysis were conducted during the summer and fall of 1979 using a combination of techniques, including air photo interpretation, air and ground reconnaissance, quantitative transect sampling, and computerized data reduction (Cronan and DesMeules, 1985). At the outset, air photographs (B&W and IR) were used to stratify each watershed into major forest cover types. In the field, each major cover type was first surveyed qualitatively for species composition, general successional status, and disturbance history. A series of random transect surveys were then run to develop quantitative estimates of forest tree structure in each cover type. These transect lines sampled a complete range of topographic, soil, and geologic conditions.

Tree species were sampled using 9 m wide belt transects oriented along specific compass bearings. In each transect, all tree species with diameters (dbh) ≥ 2.5 cm were identified to species, measured for dbh to the nearest 0.1 cm with a metal caliper, and recorded as either living or dead.

The transect data were used to compute estimates by species for each of the following quantitative stand parameters: live tree density, live tree frequency, live tree basal area, relative density, relative frequency, relative basal area, and importance value (Whittaker,

1970). In addition, estimates of frequency, density, and basal area were made for total living and dead trees. For the final synthesis, the quantitative forest vegetation data were condensed to five major forest cover types: hardwood, mixed, lowland conifer, logged hardwood, and montane conifer. These cover types were delineated using the following guidelines: (1) in hardwood communities, at least two-thirds of the dominance (I.V.) was represented by hardwood species; (2) in coniferous communities, at least two-thirds of the dominance was coniferous; (3) mixed communities fell between the bounds for the hardwood and coniferous types; and (4) the logged hardwood cover type was characterized by extensive logging disturbance within the last ten years. Lowland and montane coniferous cover types were distinguished by differences in species composition and elevation.

Vegetation maps were prepared using air photo interpretations, field survey notes, and transect data to delineate cover type boundaries on air photo acetate overlays. Each map was measured with a planimeter to determine the percent areal coverage by each vegetation cover type.

For the final stage of analysis, the vegetation data were used to develop weighted tree species dominance rankings for each watershed and to produce watershed-level estimates of tree canopy leaf area, forest aboveground biomass, and forest net aerial productivity. These and other techniques used in the vegetation study are described in detail in a companion paper by Cronan and DesMeules (1985).

3.2. Soil chemistry

Soil pits were excavated at five of the lysimeter sampling stations in the Panther and Woods Lake watersheds. Because of time limitations, Sagamore watershed was not included in the soil chemistry sampling. At each soil pit, duplicate soil samples were collected by horizon for chemical analysis. Samples were air-dried, were sieved to pass a 2 mm brass sieve, and were analyzed for the following chemical parameters. Soil pH was measured in a 1 : 1 soil/water paste using a Fisher Accumet pH meter. Organic matter was estimated by combustion at 550 °C for 4 hr. Exchangeable cations were extracted with $1 N$ ammonium acetate at pH 7 and were analyzed by flame AAS. Exchangeable acidity was estimated by extraction with $1 N$ KCl, followed by titration of the supernatant with NaOH to a phenolphthalein endpoint. Extractable SO_4^{2-} and NO_3^- were determined by extraction with $0.01 M NaH_2PO_4$, coupled with ion chromatographic analysis.

3.3. Solution chemistry

Soil solution samples were collected in the unsaturated zone with ceramic porous cup tension lysimeters (Soil Moisture Corp.) that were installed at depths of 20 cm (O/E horizon interface) and 50 cm (lower B horizon) in each watershed. Before installation, the lysimeters were thoroughly cleaned by leaching with 10% HCl, followed by multiple rinses with distilled water. Tests indicated that these lysimeters leached negligible amounts of aluminum after the cleaning procedure. Ten pairs of shallow and deep lysimeters were placed in each watershed during September 1979 and were allowed to equilibrate until February 1980. These lysimeter sampling locations were selected to include a range of vegetation types and topographic positions in each watershed.

Soil solution samples were collected every 2 to 4 weeks between February 1980 to October 1981, except during the mid-winter period when the lysimeters were usually frozen. For sampling soil solutions, the lysimeters were pumped dry, were evacuated with 0.3 bar tension for 24 hr, and the lysimeter contents were then pumped into clean glass and polyethylene sample bottles. At the same time, additional samples were collected from a number of first order streams in each watershed and from the inlets and outlets of Panther, Sagamore, and Woods lakes. Following collection, the samples were placed on ice in the dark and were transported to the laboratory for processing. Samples were processed and analyzed for inorganic and organic constituents using AAS, modified Technicon colorimetric methods, and infrared carbon analysis as described by Cronan *et al.* (1978), Cronan (1979), and Cronan and Aiken (1985). Monomeric Al measurements were made on unfiltered samples by Dr C. L. Schofield (Cornell University) using a modified ferron-orthophenanthroline method described by Driscoll (1980) and Schofield *et al.* (1985).

4. Results and Discussion

4.1. FOREST STRUCTURE

The forest vegetation analysis revealed that Panther, Sagamore, and Woods watersheds all contained the same general northern hardwoods tree flora, but that the watersheds exhibited several differences in terms of relative tree species composition and forest tree community structure (Cronan and DesMeules, 1985). General cover type distributions for each catchment are shown on the vegetation maps in Figure 1a-c. These maps illustrate that all three watersheds were dominated by various mixtures of hardwood vegetation, with additional amounts of coniferous cover ranging from roughly 5% of the watershed at Woods Lake to almost 30% of the catchment at Sagamore Lake. The quantitative differences in forest community structure between the three ILWAS watersheds are presented in Tables I and II and are discussed below.

4.1.1. *Panther Watershed*

Panther watershed included 75% hardwood forest, 14% mixed forest, 10% lowland coniferous forest, and 1% wetland cover. The hardwood community averaged 1100 stems ha^{-1} and showed a mean basal area of 20.7 m^2 ha^{-1}. As was the case in Woods and Sagamore watersheds, the Panther hardwood community was heavily dominated by beech (*Fagus grandifolia*) and maple (*Acer saccharum*). Subdominant species included red spruce (*Picea rubens*), yellow birch (*Betula alleghaniensis*), eastern hemlock (*Tsuga canadensis*), red maple (*Acer rubrum*), and striped maple (*Acer pensylvanicum*).

4.1.2. *Woods Watershed*

Woods Lake catchment contained 88% hardwood forest, 5% lowland coniferous forest, 3% mixed forest, and 4% wetland cover. The hardwood community had a mean

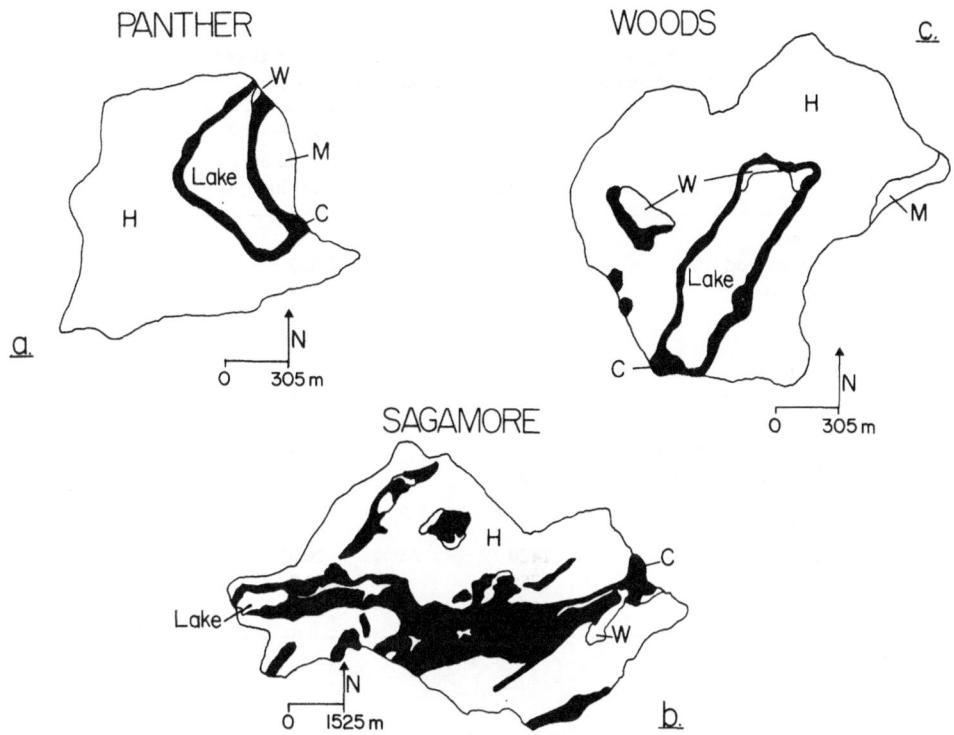

Fig. 1. General forest cover maps for Panther, Sagamore, and Woods Lake watersheds. Cover types include hardwood forest (H), coniferous forest (C), mixed forest (M), and herb/shrub dominated wetland (W). In the Sagamore map, the small percentage of logged hardwood was mapped with the general hardwood type, and the small percentage of montane conifer was combined with the overall coniferous cover type. (After Cronan and DesMeules 1985).

density of 1400 stems ha^{-1} and an average basal area of 20.9 m^2 ha^{-1}. Like the other two watersheds, the Woods hardwood community was dominated by beech and maple, although in this case, red maple was the important co-dominant. This pronounced role for red maple in the Woods Lake hardwood community appeared to correlate with wetter soils, a high water table, and a younger forest in this particular upland community.

4.1.3. Sagamore Watershed

The vegetative cover in Sagamore watershed was composed of approximately 57% hardwood forest, 26% lowland coniferous forest, 6% logged hardwood forest, 2% montane coniferous forest, and 9% wetland cover. As such, the large and complex Sagamore watershed appeared to be a reasonable cross-section of the full range of vegetation cover types in the Adirondack Park. The Sagamore catchment was particularly intriguing from the standpoint of forest age structure. Within this 48 km^2 watershed, stands ranged in age from recent regrowth associated with logging and windthrow disturbance to old-age hardwood and coniferous stands containing red spruce, white pine, sugar maple, and yellow birch trees that were 200 to 300 yr old. The Sagamore

TABLE I

Comparison of average density and basal area characteristics between the
Panther, Sagamore, and Woods Lake Watersheds.

Watershed	Average density (stems ha^{-1})	Average basal area (m^2 ha^{-1})
Panther		
Hardwood	1120	20.7
Mixed	1810	26.4
Conifer	2850	30.8
Sagamore		
Hardwood	1460	30.3
Conifer	2180	33.6
Logged Hardwood	660	9.4
Montane Conifer	3350	29.5
Woods		
Hardwood	1420	20.9
Conifer	1780	23.2
Mixed	2180	37.5

TABLE II

Summary of major forest vegetation parameters in the tree strata of the ILWAS watersheds

Parameter	Panther	Sagamore	Woods
Basal area, m^2 ha^{-1}			
Live	22.4	29.9	21.6
Dead	7.0	7.8	4.1
Density, stems ha^{-1}			
Live	1400	1700	1500
Dead	190	560	250
Biomass MT ha^{-1}	154	199	143
Net production, MT ha^{-1} yr^{-1}	6.4	8.8	7.6
Annual biomass increment (%)	4.1	4.4	5.3
Leaf area, m^2 ha^{-1}	58 400	71 950	51 900
Leaf area index	5.8	7.2	5.2
Forest cover comparisons			
% Hardwood	75	57	88
% Mixed	14	–	3
% Conifer	10	29	5
Leading dominants	Beech	Beech	Beech
	Red spruce	Red spruce	Red spruce
	Sugar maple	Sugar maple	Sugar maple
			Yellow birch

hardwood community, itself, exhibited the largest density and basal area values for living trees of any of the three ILWAS watersheds, averaging 1500 stems ha^{-1} and 30.3 m^2 ha^{-1}, respectively. Species composition in the hardwood forest was generally similar to the Woods and Panther hardwood forests, and was dominated by American beech and large specimens of sugar maple. Subdominant tree species included: red spruce, red maple, striped maple, and yellow birch. The lowland coniferous forest in Sagamore watershed was typical spruce-fir vegetation dominated by red spruce and balsam fir (*Abies balsamea*).

4.1.4. *Biomass-Productivity-Leaf Area Projections*

The quantitative belt transect data were used to produce estimates of tree biomass, net aerial productivity, and canopy leaf area for each ILWAS watershed (Cronan and DesMeules, 1985). These derived values indicated that aboveground biomass in the ILWAS catchments decreased in the following order: Sagamore (199 MT ha^{-1}) > Panther (154 MT ha^{-2}) > Woods (143 MT ha^{-1}). For comparison, biomass at Hubbard Brook Experimental Forest in New Hampshire ranges from 102 to 162 MT ha^{-1} (Whittaker *et al.*, 1974). Aboveground net productivity at the ILWAS sites decreased in the following order: Sagamore (8.8 MT ha^{-1} yr^{-1}) > Woods (7.6 MT ha^{-1} yr^{-1}) > Panther (6.4 MT ha^{-1} yr^{-1}). Using these figures and expressing net productivity as a percentage of standing biomass, it is apparent that the forest communities in Woods watershed have the fastest annual rate of growth (5.3% yr^{-1}), followed by Sagamore watershed (4.4% yr^{-1}) and Panther watershed (4.1% yr^{-1}). This conclusion agrees with the general field observation that Woods watershed forest is the youngest and therefore most rapidly growing of the three ILWAS systems. Leaf area projections indicated that leaf area indices (LAI) in the ILWAS watersheds ranged from approximately 5.2 in Woods catchment to 7.2 in Sagamore. As such, these data suggested that Sagamore watershed, with its high percentage of conifers, higher LAI, and greater elevational range, may experience significantly greater inputs of dry-deposited and horizontally-impacted pollutants such as SO_2 and $(NH_4)_2SO_4$ (Lovett *et al.*, 1982). This phenomenon could then contribute to higher sulfate concentrations in Sagamore surface waters as compared to Panther and Woods watersheds.

4.1.5. *Inter-Watershed Comparisons*

The forests in each watershed were also compared using the composite relative dominance ranking shown in Table III. With this ranking, it was possible to compare the weighted overall importance of each tree species between watersheds. As indicated, American beech was the consistent dominant tree species in each of the three ILWAS catchments, with 25 to 48% of the forest community dominance. At the same time, sugar maple was consistently the third most abundant species in each watershed, with 11 to 13% of the total dominance. However, the second most dominant species (with 22 to 23% dominance) in Panther and Sagamore watersheds was red spruce, while Woods watershed showed strong dominance (23%) by red maple. Thus, in terms of species importance, Panther and Sagamore were dominated by beech and spruce, whereas

TABLE III

Comparison of the relative dominance of tree species in the ILWAS watersheds. This index expresses the percent of the total forest dominance attributable to a given species in each watershed.

Species	Panther (%)	Sagamore (%)	Woods (%)
American beech	48.4	25.2	27.2
Red spruce	21.6	22.7	9.7
Sugar maple	12.7	12.9	11.3
Red maple	1.6	6.9	22.9
Yellow birch	8.2	6.0	11.3
Balsam fir	0.02	7.1	0.6
Striped maple	0.8	4.9	11.2
Eastern hemlock	5.3	–	–
White cedar	–	2.5	–

Woods was dominated by beech and maple. On a broader scale, the rankings showed that 90% of the dominance in all three ILWAS watersheds was attributable to beech-maple-birch-spruce forest cover.

Overall, the three ILWAS watersheds can be viewed as three closely related variants of the northern hardwood-spruce-fir complex of the Adirondack region. Results showed that the most acid system (Woods) and the least acid system (Panther) were relatively similar with respect to forest composition and structure. The major differences between these two systems related to the apparently younger age of the Woods Lake forest, the small proportion of conifers at Woods Lake, and the relatively large beaver meadow in the Woods watershed. In comparison with Woods and Panther, Sagamore watershed presented several interesting contrasts. First, the forest cover seemed on average to be older and larger than forests in the other two catchments. Thus, the forest exhibited appreciably larger values for biomass, basal area, and canopy leaf area. The large percentage coniferous cover in Sagamore watershed was distinctive and probably influenced watershed biogeochemistry in at least two ways. As suggested earlier, the extensive coniferous cover and associated large LAI in Sagamore probably increased the total loading of SO_x and other atmospheric pollutants to this lake-watershed system. In addition, the biogeochemistry of dissolved organic C in this brown-water lake was strongly influenced by the extensive areas of riparian spruce-fir forest associated with high water table conditions (Cronan and DesMeules, 1985; Cronan and Aiken, 1985).

4.2. SOIL CHEMISTRY IN THE ILWAS WATERSHEDS

The soils in the ILWAS watersheds were predominantly Becket series Haplorthods, Fragiorthods, and Haplaquods. For reference, a typical profile description is shown in Table IV. In cross-section (Figure 2), the soil stratigraphy in these watersheds was characterized by a highly organic forest floor horizon, a heavily leached E horizon (3% organic matter), and a Bhs spodic horizon that was significantly enriched in soil organic

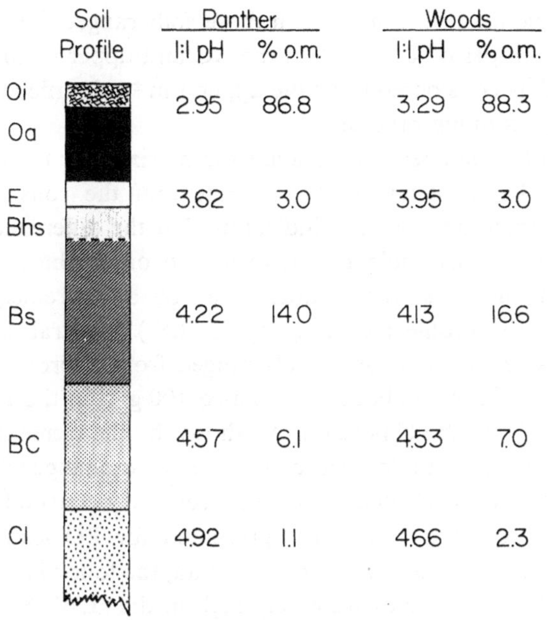

Soil Profile	Panther		Woods	
	1:1 pH	% o.m.	1:1 pH	% o.m.
Oi / Oa	2.95	86.8	3.29	88.3
E / Bhs	3.62	3.0	3.95	3.0
Bs	4.22	14.0	4.13	16.6
BC	4.57	6.1	4.53	7.0
Cl	4.92	1.1	4.66	2.3

Fig. 2.   Soil profile patterns for pH and organic matter in the Panther and Woods catchments. For reference, the E horizon is 4 cm thick.

TABLE IV

Illustrative soil profile description from Panther watershed pit 1. This Becket series Fragiorthod was described by the Soil Conservation Service in the Town of Webb, Herkimer County, New York, at 74°55′18″ W and 43°41′48″ N. The profile varies from a typical Fragiorthod in that it has 17 cm of O horizon over mineral soil.

O_i	17–12 cm	Black (5 2/1) hemic material; many partially decayed leaves and conifer twigs and needles.
O_a	12–0 cm	Black (5 2/1) sapric material; unidentified as to origin. pH 3.25
E	0–4 cm	Reddish gray (5 5/2) loamy sand; massive single grain; very friable. pH 3.77
Bhs	4–9 cm	Reddish black (10 2/1) smeary loam; weak medium subangular blocky structure; very friable; few stones; common fine and medium roots; clear smooth boundary. pH 3.97
Bs	9–20 cm	Dark reddish brown (2.5 3/4) smeary loam; weak medium subangular blocky structure; very friable; few stones; common fine and medium roots; clear smooth boundary. pH 4.35
BC	20–40 cm	Dark brown (7.5 4/4) sandy loam; weak coarse platy structure; friable; few stones; very few roots; clear smooth boundary. pH 4.66
Cl	40–55 cm	Dark brown (10 4/3) very fine sand; massive to weak coarse platy structure; friable; 5% coarse fragments; no roots; clear broken boundary. pH 4.80
C2	55⁺ cm	Dark brown (10 3/3) gravelly loamy sand; with few faint fine to medium dark brown (7.5 4/4) mottles; medium platy; very firm weakening with depth to firm; 20% coarse fragments; no roots.

Note: soil colors are all YR.

matter (15% organic matter). Soil pH's in these soils ranged from a low of 2.9 to 3.3 in the O horizon to a pH of 4.5 to 4.9 in the BC and upper C horizons. It should be noted that the 'soil' is considered to be the upper 1 m ± of surficial material which has been altered by soil-forming processes.

In Tables V to VII, soil chemical data are summarized for the Panther and Woods Lake watersheds. These mean values illustrate that the soils in these two study catchments were remarkably similar, indicating that the differences in lake alkalinity between these two systems could not be explained on the basis of contrasts in soil chemistry. Overall, the soils were characterized by an extremely low percent base saturation through the profile (generally $< 8\%$ B.S.). Neutral salt cation exchange capacity (CEC) values in these Spodosols ranged from approximately 25 to 30 meq 100 g^{-1} in the forest floor, to about 7 to 20 meq 100 g^{-1} in the Bs horizon, to a low of 1 to 2 meq 100 g^{-1} in the C horizon. As shown by the element ratios in Table VI, roughly 75% of the base saturation was contributed by exchangeable Ca. The table also indicates that exchangeable element ratios were relatively constant through the profile.

In Table VII, representative data are presented for PO_4-extractable NO_3^- and SO_4^{2-} in the ILWAS soils. Based upon these data, the following points can be made: (1) extractable soil NO_3^- values were very high in the ILWAS Spodosols (11.1 keq

TABLE V

Summary soil chemistry data for the ILWAS watersheds.

Horizon	Exchangeable (meq 100 g^{-1} dry soil)						
	Ca	Mg	K	Na	Acidity	CEC	% B.S.
Panther[a]							
O	8.304	2.175	0.826	0.055	15.581	26.941	42
E	0.169	0.041	0.040	0.010	3.500	3.760	7
Bh	1.008	0.095	0.056	0.009	13.581	14.749	8
Bhs	0.903	0.074	0.066	0.015	16.433	17.491	6
Bs	0.227	0.037	0.036	0.007	7.670	7.977	4
BC	0.168	0.031	0.022	0.006	2.272	2.499	9
C	0.114	0.020	0.042	0.017	0.965	1.158	17
Woods							
O	8.316	1.703	0.576	0.040	14.032	24.667	43
A	0.948	0.190	0.196	0.019	26.171	27.524	5
E	0.184	0.040	0.040	0.003	6.198	6.465	4
Bh	0.216	0.050	0.063	0.006	19.292	19.627	2
Bhs	0.293	0.059	0.054	0.009	9.957	10.372	4
Bs	0.158	0.032	0.029	0.012	7.404	7.635	3
BC	0.071	0.013	0.013	0.008	4.378	4.483	2
C	0.059	0.008	0.004	0.007	2.082	2.160	4

[a] Exchange acidity includes H^+ and Al.

CEC = sum of exchangeable cations and acidity.
These mean values are based on five soil pits per watershed.

TABLE VI

Relative ion ratios for exchangeable cations in the ILWAS soils. Each element is expressed as a percentage of the sum of equivalents of cations.

Horizon	Ca	Mg	K	Na
Panther				
O	73	19	7	1
E	65	16	15	4
Bh	86	8	5	1
Bhs	85	7	6	2
Bs	74	12	12	2
BC	74	14	10	2
C	59	10	22	9
Woods				
O	78	16	5	1
A	70	14	14	2
E	69	15	15	1
Bh	64	15	19	2
Bhs	71	14	13	2
Bs	68	14	13	5
BC	68	12	12	8
C	76	10	5	9

TABLE VII

Extractable SO_4^{2-} and NO_3^- in a representative soil profile from the ILWAS watersheds. Panther pit 1 was extracted with 0.01 M NaH_2PO_4 to estimate total soluble plus extractable ions.

Horizon	SO_4^{2-}	NO_3^-
	Meq 100 g^{-1} soil	
O	0.360	1.330
E	0.010	0.100
Bh	0.120	0.450
Bhs	0.170	0.550
Bs	0.160	0.330
BC	0.060	0.030
C	0.040	–
Estimated Storage in upper 35 cm of soil profile	3.6 keq ha^{-1} 57 kg S ha^{-1}	11.1 keq ha^{-1} 156 kg N ha^{-2}

ha^{-1}) and were much higher than extractable SO_4^{2-} equivalents (3.6 keq ha^{-1}); (2) storage of extractable SO_4-S in the upper 35 cm of forest floor and soil (57 kg S ha^{-1}) was equivalent to approximately 2 to 4 yr of atmospheric deposition; and (3) storage of extractable NO_3-N in the upper 35 cm of the soil profile (156 kg N ha^{-1}) was equivalent to roughly 10 to 20 yr of atmospheric deposition.

As a final component to the soil chemistry research effort, a laboratory experiment was conducted to examine the 'enigma' of the alkalinity differences in the ILWAS lakes. During the initial phases of the ILWAS study, it was thought that the circumneutral waters in Panther Lake resulted from the influence of a calcareous deposit in the basin. Later, data were collected indicating that the catchment contains virtually carbonate materials (April and Newton, 1985). With this the case, attention focused on a hypothesis suggesting that Panther Lake is circumneutral because it is fed predominately by ground water that has experienced a long residence time in the soil and till. Given this long residence time, one might predict that weathering reactions could proceed toward completion, resulting in elevated solution pH's and alkalinities. However, this hypothesis assumes that the soil and/or till materials are thermodynamically capable of producing an equilibrium solution pH around 7. To test this underlying assumption, a four-week neutralization kinetics experiment was performed with separate soil horizons from Woods Lake basin. As shown in Figure 3, initial results indicated that the

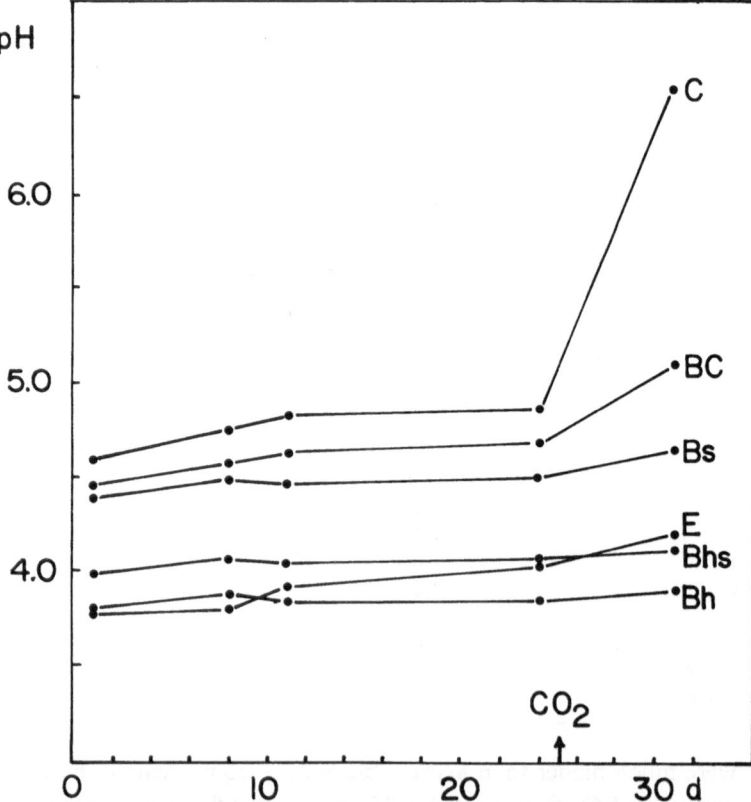

Fig. 3. A test of neutralization kinetics for soil and till materials from Woods Lake basin. Points show supernatant solution pH values over time for 1 : 1 slurries of soil horizons mixed with pH 4.0 H_2SO_4. The slurries were shaken continuously, and were centrifuged before each measurement. After CO_2 addition and incubation, the supernatants were purged back to an air-equilibrated pH (see arrow for CO_2 addition).

neutralization of acid deposition by the soil or till is time-dependent and relatively slow. Even after three weeks, none of the supernatant pH's had approached pH 7. To accelerate the experiment, the decision was made to produce an extreme simulation of the carbonation reactions that naturally occur with CO_2 – enriched ground water (by purging with CO_2, followed by re-equilibration with air). As shown in Figure 3, that treatment was sufficient to produce a pH of 6.58 in the C horizon (upper till) supernatant, with somewhat less dramatic results in the other soil horizons. Overall, these results were interpreted as follows: (1) given enough time and CO_2, the glacial till could be expected to react with acid deposition percolate in such a way as to produce a circum-neutral solution pH and substantial amounts of alkalinity; and (2) because of the aluminum and/or organic buffering systems present in the O, A, and B horizons, it is unlikely that these soil horizons could be expected to yield enough alkalinity to produce the solution pH's observed in Panther Lake (even if the water residence time was very long).

4.3. SOLUTION CHEMISTRY AND ION TRANSPORT

Results from the solution chemistry investigation indicated that soil water chemistries were generally indistinguishable between Woods and Panther catchments, but that significant differences in water chemistry occurred between different soil horizons, seasons, and vegetation types (Table VIII). Soil water from the O/E horizon interface in the unsaturated zone typically contained high concentrations of dissolved organic C (20 to 80 mg C L^{-1}), exhibited a mean pH around 3.8, and was dominated by H_2SO_4, HNO_3, and organic acids. Aqueous Al in this zone was generally undersaturated with respect to mineral solid phases and included a large proportion of organically complexed Al (Al_a = 600 to 800 µg L^{-1} and $\%Al_0$ = 30 to 80%). Mean alkalinity ($C_B - C_A$) values for these shallow soil solutions ranged from -54 to -89 µeq L^{-1}.

Soil solutions from the lower Bs horizon (50 cm depth) in the unsaturated zone were distinctly different from nearsurface soil water. These solutions were characterized by high concentrations of aqueous Al (Al_a = 1000 to 1300 µg L^{-1}), exhibited a mean pH around 4.5, had much lower DOC concentrations (5 to 7 mg C L^{-1}), and contained substantial concentrations of SO_4^{2-} and NO_3^-. Calculated alkalinity values for these soil solutions were higher than those observed in the shallow soil water, ranging from -11 to -37 µeq L^{-1}. Much of this change in alkalinity resulted from a decrease in SO_4^{2-} concentration between upper and lower horizons and from the combined effects of Al mobilization and H^+ consumption between soil zones.

Lake inlet and outlet chemistries are also presented in Table VIII. These data illustrate the marked contrasts in solution pH, Ca concentration, and alkalinity between the three ILWAS systems. The data for Woods and Panther watersheds also indicate the large decrease in SO_4^{2-} and NO_3^- concentrations observed between the deeper soil horizons and surface waters.

Overall results showed that SO_4^{2-} from atmospheric deposition was the dominant anion in soil waters from all three catchments and in the surface waters of both Woods and Sagamore watersheds. Panther was the only catchment which generated enough

TABLE VIII

Comparative major ion chemistries for drainage waters in the three ILWAS watersheds during the summer (June-October) and winter (November-May) periods of 1980 and 1981. Summer data are designated 'S' while winter periods are designated 'W'. Water chemistry data for each season and watershed include tension lysimeter soil solutions at 20 cm and 50 cm depths, lake inlets, and lake outlets. All data except Al_o are presented in ueq L^{-1}. Al_o data are expressed as a percent of Al_a.

Site		pH	Ca	Mg	K	Na	NH_4^+	Al_a	Al_o	Fe	Sum (+)	SO_4^{2-}	NO_3^-	Cl^-	F^-	Sum (−)	DOC
Panther																	
S	20 cm	3.82	139	41	28	23	2	69	58%	12	465	210	81	32	2	325	32
S	50 cm	4.54	98	33	9	27	1	106	13%	–	303	151	97	20	5	273	7
S	Inlet	7.22	186	68	12	58	1	0	–	–	325	116	45	10	–	171	3
S	Outlet	7.33	208	54	12	43	2	0	–	–	319	122	8	12	–	142	4
W	20 cm	3.90	116	30	36	17	6	64	45%	12	407	147	153	30	2	332	22
W	50 cm	4.55	97	36	9	25	3	129	11%	–	327	138	163	16	5	322	7
W	Inlet	6.94	175	66	13	47	1	0	–	–	302	112	69	9	–	190	3
W	Outlet	6.98	190	49	12	38	2	0	–	–	291	116	37	11	–	164	4
Woods																	
S	20 cm	3.87	108	31	28	17	2	88	34%	9	418	210	90	31	2	333	27
S	50 cm	4.50	84	21	7	28	2	102	12%	–	276	191	28	19	5	243	7
S	Inlet	4.91	77	22	4	30	4	10	36%	–	159	137	7	14	–	158	7
S	Outlet	4.88	66	21	6	20	4	9	1%	–	139	121	10	10	–	141	2
W	20 cm	3.93	103	27	39	15	8	67	29%	9	386	150	116	26	2	294	20
W	50 cm	4.50	83	22	7	27	2	125	11%	–	298	169	76	17	5	267	7
W	Inlet	4.59	84	22	7	27	2	52	19%	–	220	141	40	11	–	192	7
W	Outlet	4.78	70	18	7	20	3	23	8%	–	158	119	36	11	–	166	2
Sagamore																	
S	20 cm	4.16	145	43	40	27	8	47	60%	–	379	193	56	27	–	276	21
S	50 cm	4.93	160	68	27	43	7	24	18%	–	341	150	66	22	–	238	5
S	Inlet	6.02	151	55	10	43	1	2	–	–	263	177	7	19	–	203	8
S	Outlet	6.14	138	56	13	39	1	1	–	–	249	172	14	17	–	203	7
W	20 cm	4.20	146	43	67	32	17	52	50%	–	420	184	127	31	–	342	21
W	50 cm	4.94	136	54	30	35	2	25	36%	–	294	144	59	21	–	224	7
W	Inlet	5.10	135	47	17	38	5	14	41%	–	264	174	32	17	–	223	8
W	Outlet	5.75	140	54	15	35	2	3	–	–	251	163	40	16	–	219	7

DOC = units are mg C L^{-1}

Ala = Total monomeric Al

Al_o = Organically complexed monomeric Al

The anion deficit corresponds to organic ligands for most samples below pH 5.

alkalinity (through carbonic acid weathering and ion exchange in the groundwater zone) to produce bicarbonate dominance in the lake anion chemistry. Soil-derived organic and nitric acids were important primarily in the soil zone and generally did not contribute significantly toward acid transport to surface waters. However, during periods of high flow and low vegetative uptake, the importance of NO_3^- (and to some extent organic anions) increased significantly in some ILWAS surface waters.

It is important to note that mean NO_3^- concentrations in the ILWAS soil solutions were unusually high, ranging up to 3 × the concentration in wet deposition. This pattern can be contrasted with other forested watersheds in the Northeast where NO_3^- values are near detection limits during the growing season (Vitousek, 1977; Likens *et al.*, 1977; Cronan, 1980). The data suggest either that the ILWAS systems have become N-saturated or that plant uptake of NO_3^- has changed (e.g. it has been hypothesized that excess NH_4^+ or aqueous Al can interfere with NO_3^- uptake). Reports from West Germany (E. Matzner, pers. comm.) indicate that a similar pattern of net NO_3^- leaching has recently begun in the Solling Forest. Overall, the ILWAS data suggest that further increases in N deposition may exceed N requirements in the terrestrial system and may lead to increased export of nutrient catons and aqueous Al to surface waters.

In a more detailed companion study on humic chemistry, Cronan and Aiken (1985) showed that soil-derived organic acids exerted important effects on solution acidity and metal transport in the ILWAS systems. Results indicated that organic ligands accounted for approximately 25% of the total anion equivalents in shallow soil solution samples and 10% of the anion equivalents in deeper soil solutions. The majority of DOC in these natural waters was acidic in nature, so that as DOC increased in the low akalinity interflow drainage from these watersheds, solution acidity tended to increase. However, overall transport of organic acidity to surface waters was relatively limited in all three watersheds.

There was a substantial mobilization and transport of aqueous Al in the ILWAS soils. In the upper soil profile, Al_a (total monomeric Al) concentrations averaged 500 to 800 μg L^{-1} and there was a strong influence by soil-derived organic acids on Al speciation. Depending upon hydrologic conditions, this upper soil water would typically either move laterally to contribute to storm flow or move vertically into the B horizon. In the B horizon, mean Al_a concentrations increased 2 × to a range of 1000 to 1300 μg L^{-1} and the speciation shifted almost exclusively toward inorganic monomeric Al. This dramatic eluviation of inorganic Al from the Bs horizon appeared to be strongly influenced by H_2SO_4 and HNO_3 derived largely from atmospheric deposition. During most of the year, Al leached from the B horizon was immobilized at some point between the lower soil profile and the lake feeder streams. In winter, this immobilization was not so pronounced, permitting a major flux of aqueous Al to systems like Woods Lake (Table VIII). Based upon solubility calculations and observed concentration trends, the data suggest that Al immobilization in the ILWAS systems may be explained by the precipitation of an inorganic basic aluminum sulfate like basaluminite ($Al_4(OH)_{10}SO_4$).

The combined results from this study component and from Peters and Murdoch (1985) and April and Newton (1985) showed that the differences in lake pH and alkalinity between Woods and Panther watersheds could not be explained on the basis of contrasts in the solution chemistry of any given soil stratum. Instead, the differences appeared to stem from the relative amounts of water moving laterally into surface waters from the different horizons. For example, the chemistry of Woods Lake indicated a predominance of shallow lateral flow paths and a large contribution from B horizon soil water. This was particularly true during the winter (Table VIII) when solution chemistries for the B horizon and lake inlet were identical in several respects. In contrast, the chemistry of Panther Lake reflected the influence of deeper groundwater contributions characterized by higher Ca and HCO_3^- concentrations.

5. Summary

The ILWAS watersheds in the Adirondack Region of New York State contain northern hardwood – spruce – fir forest vegetation and are dominated by American beech, sugar maple, red spruce, red maple, and yellow birch. On an areal basis, the watersheds include 57 to 88% hardwood forest, while coniferous cover in the three catchments ranges from Sagamore (28%) > Panther (10%) > Woods (5%). Above-ground tree biomass estimates for the ILWAS catchments decrease in the following order: Sagamore (199 MT ha^{-1}) > Panther (154 MT ha^{-1}) > Woods (143 MT ha^{-1}). Forest growth rate estimates are 5.3, 4.4, and 4.1% yr^{-1} for Woods, Sagamore, and Panther, respectively. Average leaf area indices (LAI) for the watersheds range from a low of 5.2 in Woods to a high of 7.2 in Sagamore. The higher leaf area in Sagamore may enhance the collection of dry deposition, causing higher SO_4^{2-} concentrations in runoff from the watershed.

Soils in the ILWAS watersheds are predominantly Becket series Haplorthods, Fragiorthods, and Haplaquods that have developed under the intense weathering influence of organic cheluviation and podzolization. Soil chemical properties in the upper 1 m of surficial material are almost identical in the Woods and Panther watersheds. The soils are characterized by a very low percent base saturation (generally < 8% B.S.), exhibit pH values ranging from 2.9 in the forest floor to 4.7 in the BC horizon, and contain relatively small amounts of soluble plus adsorbed SO_4^{2-} (50 to 60 kg S in the upper 35 cm of soil profile).

Soil water chemistries in the Woods and Panther catchments are virtually indistinguishable; thus, the differences in lake pH and alkalinity between these two systems cannot be explained on the basis of contrasts in soil chemistry and soil solution chemistry. Much of the explanation for the contrasting behavior of the ILWAS lakes is related to differences in hydrologic flow paths, coupled with distinct differences in solution chemistry between separate soil horizons and till strata. In general, sulfate is the dominant anion in soil waters from all three catchments and in the surface waters of Woods and Sagamore watersheds. Both nitrate and organic acid anions are important in upper soil horizons. There is a substantial mobilization and transport of aqueous Al throughout the soil profile in the ILWAS catchments; most Al immobilization occurs after water leaves the active soil zone.

The transport of acidity to surface waters in the ILWAS systems can best be explained through a combination of the mobile anion concept and hydrologic flow path analysis. In these systems, biotic and abiotic retention mechanisms prevent significant transport of soil-derived organic and nitric acids from the soil zone to surface waters. In contrast, SO_4^{2-} derived from acidic deposition is only partially controlled by adsorption, absorption, and precipitation reactions within the soil zone. Thus, transport of acidity and aqueous monomeric Al to surface waters is largely controlled by SO_4^{2-} flux and by water flow path through chemically distinct soil and till layers. During the winter, when biological uptake is minimal, NO_3^- fluxes may also exert a strong influence upon the transport of acidity and aqueous Al to surface waters.

Acknowledgments

The author expresses his gratitude to the following people for their contributions to this investigation:George Aiken, Mark DesMeules, Karen Cronan, Richard Konz, Sandy Skibinski, Phil Thorne, and Betty Lee. The financial support of EPRI is also gratefully acknowledged.

References

April, R. H. and Newton, R. M.: 1983, *Soil Sci.* **135**, 301.
April, R. H. and Newton, R. M.: 1985, *Water, Air, and Soil Pollut.* **26**, 373 (this issue).
Braun, E. L.: 1950, *Deciduous Forests of Eastern North America*. The Free Press, N.Y.
Cronan, C. S.: 1979, *Anal. Chem.* **51**, 1333.
Cronan, C. S.: 1980, *Oikos* **34**, 272.
Cronan, C. S., Reiners, W. A., Reynolds, R. C., and Lang, G. E.: 1978, *Science* **200**, 309.
Cronan, C. S. and Aiken, G. R.: 1985, *Geochim. Cosmochim. Acta* **49**, 1697.
Cronan, C. S. and DesMeules, M. R.: 1985, *Can. J. For. Res.* **15** (in press).
Driscoll, C. T.: 1980, 'Chemical Characterization of Some Dilute Acidified Lakes and Streams in the Adirondack Region of New York State', Ph.D. Thesis, Cornell Univ.
Goldstein, R. A., Gherini, S. A., Chen, C. W., Mok, L., and Hudson, R. J. M.: 1984, *Phil. Trans. R. Soc. Lond.* **B305**, 409.
Goldstein, R. A., Chen, C. W., and Gherini, S. A.: 1985, *Water, Air, and Soil Pollut.* **26**, 327 (this issue).
Johannes, R. H., Altwicker, E. R., and Clesceri, N. L.: 1985, *Water, Air, and Soil Pollut.* **26**, 339 (this issue).
Likens, G. E., Bormann, F. H., Pierce, R. S., Eaton, J. S., and Johnson, N. M.: 1977, *Biogeochemistry of a Forested Ecosystem*, Springer-Verlag, N.Y. 146 p.
Lovett, G. M., Reiners, W. A., and Olson, R. K.: 1982, *Science* **218**, 1303.
Peters, N. E. and Murdoch, P.: 1985, *Water, Air, and Soil Pollut.* **26**, 387 (this issue).
Schofield, C. L., Galloway, J. N., and Hendrey, G. R.: 1985, *Water, Air, and Soil Pollut.* **26**, 403 (this issue).
Vitousek, P. M.: 1977, *Ecol. Monogr.* **47**, 67.
Whittaker, R. H.: 1970, *Communities and Ecosystems*, Macmillan Co., N.Y. 162 p.
Whittaker, R. H., Bormann, F. H., Likens, G. E., and Siccama, T. G.: 1974, *Ecol. Monogr.* **44**, 233.

INFLUENCE OF GEOLOGY ON LAKE ACIDIFICATION IN THE ILWAS WATERSHEDS

RICHARD APRIL

Department of Geology, Colgate University, Hamilton, NY 13346, U.S.A.

and

ROBERT NEWTON

Department of Geology, Smith College, Northampton, MA 01063, U.S.A.

(Received November 1, 1984; revised July 8, 1985)

Abstract. Three lake-watersheds in the Adirondack Mountains of New York State, underlain by similar granitic bedrock and receiving similar levels of acidic deposition, were found to have very different surface water alkalinities. The chemical differences appear to be due to differences in the unconsolidated surficial materials in the basins. Woods Lake watershed (mean lake outlet pH of 4.7) is covered by thin till with many interspersed bedrock outcrops. The thinness of these surficial deposits (average depth 2 m) limits the amount of deep percolation of water and thus contact with alkalinity-producing inorganic horizons. In contrast, Panther Lake watershed (mean lake outlet pH of 6.2) is covered by thick glacial till (average depth 24 m). Here more of the precipitation comes in contact with the alkalinity-producing materials. Sagamore Lake watershed is much larger and has areas of both thick and thin deposits and lake outlet pH values intermediate to those of Woods and Panther lakes.

The soils in all three watersheds are dominated by quartz, potassium feldspar and sodic plagioclase with minor amounts of hornblende and other heavy minerals. The dominant clay mineral is vermiculite. Chemical evidence suggests the present rate of mineral weathering is less than the long-term rate in Woods Lake watershed while in Panther, the present rate may have increased relative to the long-term rate.

1. Introduction

The Adirondack Mountains of New York receive some of the most acidic deposition in all of North America. It is also an area underlain predominantly by a granitic-type bedrock – a fact that some suggest causes lakes in the Adirondacks to be highly susceptible to acidification (Hendrey *et al.*, 1980).

In this paper, the results of the research on the geological components of ILWAS are summarized. Our evidence suggests that mineral weathering and ground water flow-paths in the glacial cover mantling Adirondack lake-watersheds largely control the process of lake acidification in this region of the United States. Our discussion will mainly focus on two of the ILWAS lake-watersheds, each of which has a different response to similar inputs of acid deposition. Woods Lake is acidic with pH values ranging from 4.4 to 5.9 and Panther Lake is circumneutral with a pH range from 6 to 7 most of the year. The lake-watersheds, located within 30 km of each other, are both underlain by hornblende granitic gneiss.

Water, Air, and Soil Pollution **26** (1985) 373–386.   0049–6979/85.15
© 1985 *by D. Reidel Publishing Company.*

2. Geology of Woods and Panther Watersheds

2.1. SURFICIAL GEOLOGY

Woods Lake hydrologic basin has an area of 2.07 km^2 and has 122 m of relief. The basin is covered predominantly by a thin, sandy glacial till with an average thickness of 2.3 m with many interspersed bedrock outcrops (Figure 1). Mantling the till is a Spodosol containing a discontinuous layer of aeolian silt up to 50 cm thick. The till is locally thicker along the northwest shore of the lake where seismic refraction lines indicated thicknesses greater than 10 m. The till is generally unsorted and unstratified and was deposited directly by the last continental glacier which retreated from the area 14 000 $\pm$ 2000 yr ago.

Panther Lake hydrologic basin has an area of 1.24 km^2 and has 174 m of relief. Unlike Woods, this basin is covered predominantly by thick glacial till, averaging 24.5 m in depth (Figure 2). Spodosols have developed on the till and average less than 1 m in profile. Seismic refraction lines indicate till thicknesses greater than 30 m in some parts of the watershed and an underlying, more compact unit at depth representing, perhaps, a deposit from an earlier glaciation.

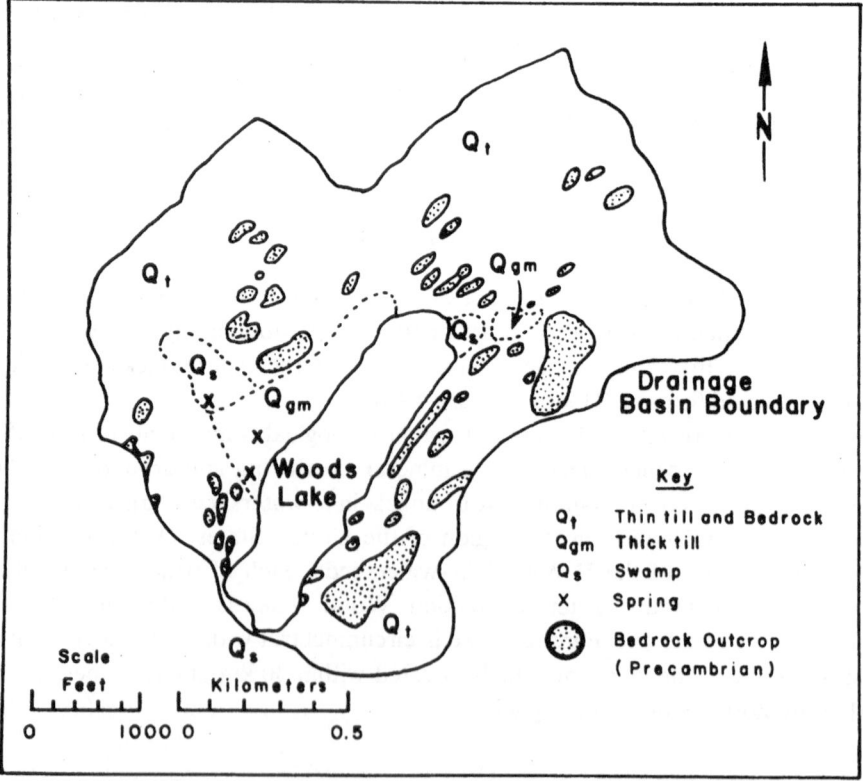

Fig. 1.   Surficial geologic map of Woods Lake basin. Average till depth is 2.3 m.

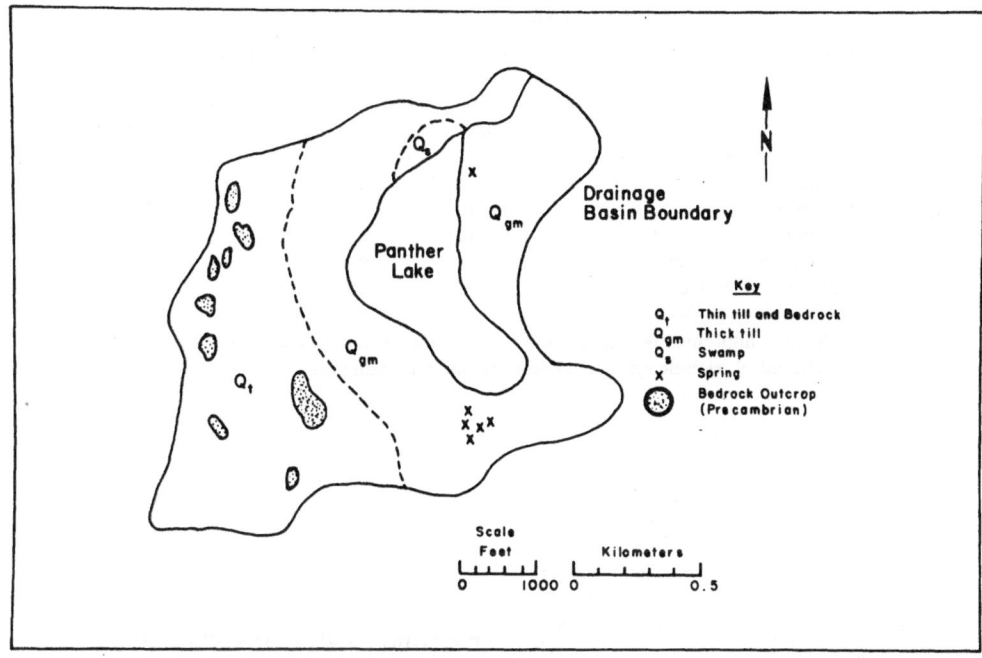

Fig. 2.   Surficial geologic map of Panther Lake basin. Average till depth is 24.5 m.

2.2. HYDROLOGIC CHARACTERISTICS

The permeability of surficial materials was determined in the field with a Tempe double tube hydraulic permeameter and in the laboratory, using a Soiltest model K-605 combination permeameter. The mean permeability of soil samples from Woods Lake watershed, measured in the laboratory on B and C soil horizons was one order of magnitude less than those from Panther Lake watershed (Table I). Both watersheds contain soil horizons having similar maximum permeabilities, but many samples from Woods watershed displayed much lower permeabilities. In addition, 23% of the samples from Woods had permeabilities too low to readily measure compared with only 9% of the samples from Panther. These differences may reflect the presence of the aeolian silt in Woods and its absence in Panther.

During the field permeability experiments, measurements of the infiltration rate of the mineral soil layers were made. Infiltration rate is initially dependent on soil moisture conditions, but as infiltration proceeds, the deficiency in soil moisture is satisfied by the infiltrating water. Eventually an equilibrium infiltration rate is achieved that is mainly a function of the physical characteristics of the soil. Figure 3 shows a comparison of typical infiltration rate curves for Woods and Panther soils. Panther Lake watershed soils had typical equilibrium infiltration rates approximately ten times those of Woods Lake soils.

TABLE I

Laboratory permeability (cm s^{-1}) of Woods and Panther soil samples
(B and C soil horizons)

	Woods	Panther
Mean	3.7×10^{-4}	5.3×10^{-3}
Minimum	9.3×10^{-6}	1.1×10^{-4}
Maximum	1.3×10^{-2}	2.2×10^{-2}
Percent Impermeable[a]	23%	9%
Number of Samples	22	15

[a] Percent impermeable represents the percent of samples analyzed which had no significant discharge from the permeameter.

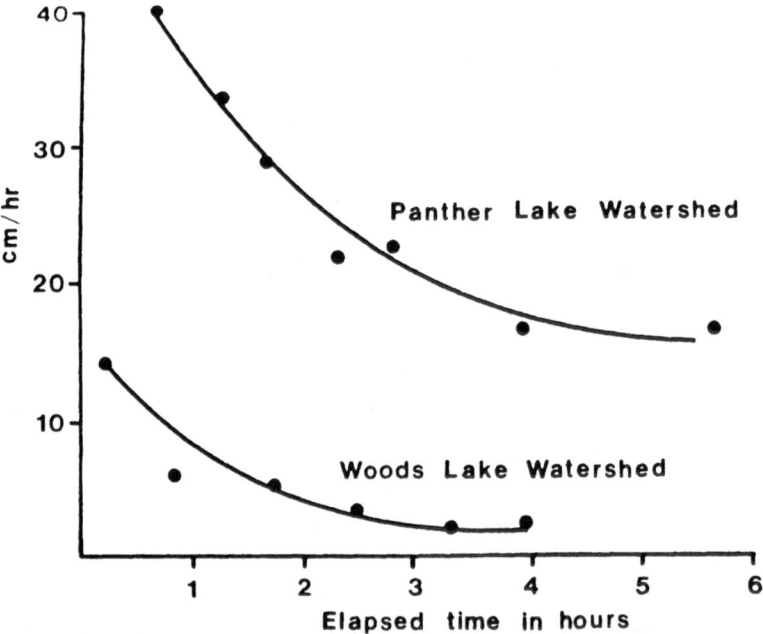

Fig. 3.   Typical infiltration rate curves for the mineral soil horizons in Woods and Panther Lake watersheds.

2.3. MINERALOGY

Approximately 20 pits were excavated in each watershed and nearly 200 soil samples were collected for analysis. Soil profiles typically developed on glacial till and average less than 1 m in depth. The bulk mineralogy of samples was determined by X-ray diffraction powder analysis, optical microscopy, and gravimetric analysis. Table II shows that the mineralogy of the till was similar in both watersheds and was dominated by quartz and feldspar with accessory hornblende, ilmenite and magnetite. If it is assumed that the soil was initially homogenous, an analysis of the current trends in the upper soil horizons shows that considerable depletion of hornblende has occurred. This

TABLE II

Average mineralogy (%)[a] of the C horizon in Woods and Panther watersheds

	Woods	Panther
Quartz	43.6	38.4
K-feldspar	31.2	31.6
Plagioclase	10.3	11.0
Altered feldspar	4.5	7.8
Hornblende	2.0	2.4
Opaque minerals	3.6	3.5
Pyroxene	0.4	0.6
Epidote	0.5	0.6
Garnet	0.6	0.4
Others	3.3	3.7
Total	100.0	100.0
Number of samples	19	24

[a] Volume percentages were determined from ribbon counts of 300 grains in each sample.

depletion, likely the result of chemical weathering since the last glaciation, was greater in the soils found in Panther watershed than those found in Woods (Figure 4). We suggest that this difference results from the disparity between the amount of water which infiltrates the soils in Woods and Panther watersheds.

The mineralogy of the sub 2-μm fraction (clay) was also determined for each sample by X-ray diffraction analysis (for methods, see April and Newton, 1983). Al-interlayered vermiculite is the dominant clay mineral in the soils with lesser amounts of illite, kaolinite and mixed-layer illite/vermiculite present. Traces of talc and possibly smectite and chlorite were observed in a few diffractograms.

2.4. SOIL GEOCHEMISTRY

Sites with similar topography, physiographic position and vegetation from each watershed were chosen for quantitative determinations of free (extractable) Fe and Al oxides/hydroxides using the methods given by Jackson (1974) and bulk and soil chemical analysis using X-ray fluorescence spectrometry. Overall, concentrations of extractable Al and Fe are lowest in A2 (E) horizons and highest in B horizons (examples shown in Table III). Electron microprobe analysis of individual hornblende grains and X-ray fluorescence analysis of hornblende separates from these soils showed average Fe concentrations (as FeO) as high as 24 weight % (Table IV). FeO values from individual grains ranged from 14 to 30.5 weight %. The extractable and bulk chemical analysis and the depletion of hornblende in the upper soil horizons strongly suggest that much of the Fe staining and accumulation in B horizons has resulted from hornblende weathering. Under coniferous stands where acidic inputs were high, enough Fe has been transported to the B horizon to form the observed cemented ortstein layer.

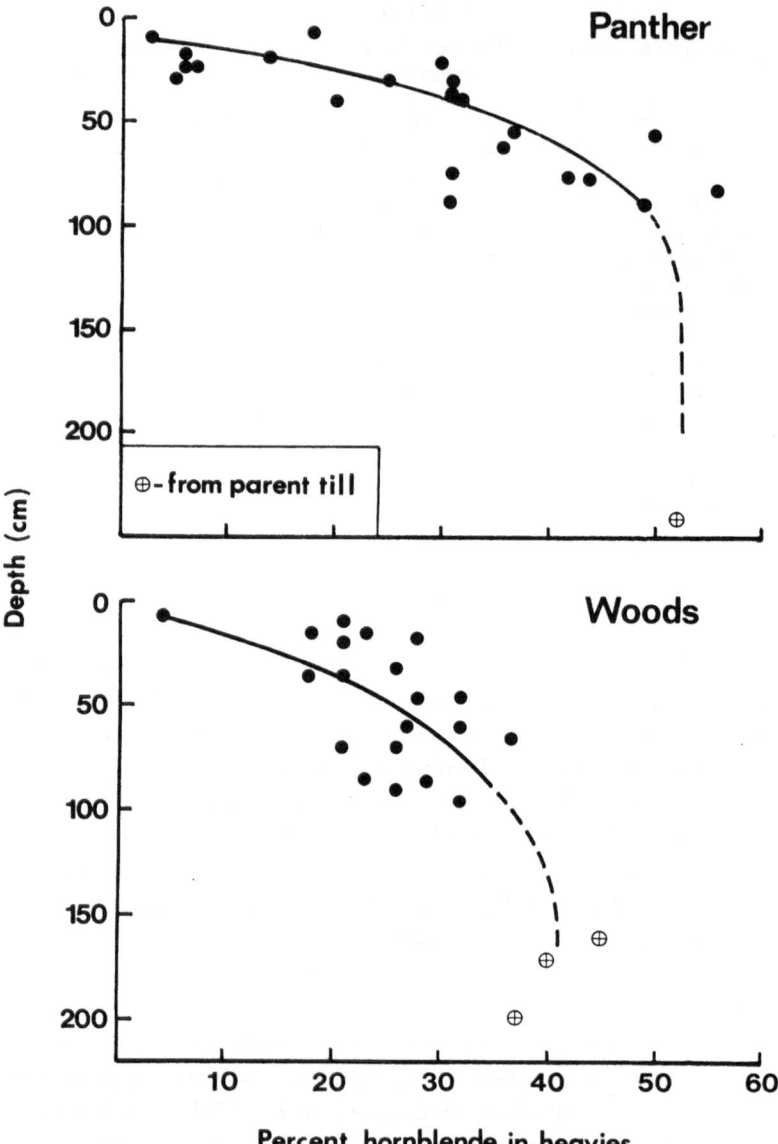

Fig. 4. Weight-% hornblende in the heavy mineral fraction versus depth in soil horizons.

As is the case for Fe, extractable Al was concentrated in B horizons. Feldspar weathering is quantitatively important in these soils and contributes to the accumulation of Al in B horizons in the form of kaolinite and as amorphous and poorly crystalline Al-oxide/hydroxide precipitates. Progressive acid dissolution experiments by Parnell (1981) on similar soils from Mt. Moosilauke, New Hampshire, indicated that extractable Al in acidic B horizons exists in four reservoirs: (1) as exchangeable Al, (2) as

TABLE III

Comparison of extractable Fe and Al with total (XRF) Fe and Al in two representative soil profiles

Soil Layer	Sample Depth (cm)	Weight-% Total	Fe Extract.	Weight-% Total	Al Extract.
Woods 115					
A2	15	n.d.[a]	n.d.	n.d.	n.d.
B22ir	35	4.95	3.32	7.26	4.12
C	60	3.17	0.83	5.71	0.73
Panther 103					
A2	15	2.29	0.97	3.74	0.10
B22ir	26	4.15	2.64	5.63	2.15
B23irm	43	2.83	1.13	6.08	1.94
B3	63	3.01	1.15	5.91	2.09
C	90	2.66	1.01	5.96	1.24

[a] n.d. – no data; insufficient sample.

Extractable Fe and Al were determined by the CBD method; total Fe and Al were determined by XRF.

TABLE IV

Average chemical composition of hornblende in Woods and Panther lake-watersheds as determined by electron microprobe and X-Ray fluorescence (XRF) analysis

Weight-% Oxide	Woods (10 grains)	Panther (15 grains)	XRF Hornblende Separate
SiO_2	39.54	39.91	42.67
MgO	5.73	5.94	7.57
CaO	10.45	10.43	11.20
Na_2O	1.98	1.99	1.17
Al_2O_3	10.86	11.29	11.07
TiO_2	2.20	2.19	2.27
K_2O	1.80	1.66	1.61
FeO^a	23.99	23.80	21.20
Cr_2O_3	0.01	0.00	–
MnO	0.35	0.40	0.32
F	0.62	0.45	–
Cl	0.43	0.53	–
Total	97.96	98.59	99.15

[a] Total Fe reported as FeO.

complexed Al, (3) as amorphous Al hydroxides, and (4) as Al interlayers in vermiculite. The extractable Al values, together with the clay-mineral data, seem to suggest that the Al-interlayered vermiculite in Adirondack soils is potentially an important Al reservoir. Whether changes in soil chemistry as a consequence of acid deposition result in vermiculite acting as a source or a sink for Al in ground water is a topic for further research.

3. Discussion

3.1. LAKE ACIDIFICATION

The sensitivity to acidification of Woods and Panther Lakes can be directly linked to differences between the surficial geology of the watersheds (Newton and April, 1982). Although the surficial deposits occupying both basins are mineralogically and texturally similar, they differ significantly in thickness and permeability. Consequently, the flow-paths along which precipitation moves enroute to the lake and the amount of acid neutralization that occurs is markedly different in each watershed.

The three basic flowpaths by which water enters a lake are: (1) direct precipitation on the lake surface, (2) overland and shallow interflow, and (3) baseflow of ground water. We have calculated the amount of water flowing along each of these flowpaths in Woods and Panther lake-watersheds using the hydrologic simulation module of the ILWAS model (see Gherini et al., 1985). Surface runoff and shallow interflow predominate in Woods Lake basin because of its thin and less permeable soils (Figure 5). Hydrologically, the watershed is a surface-water and shallow interflow-dominated system producing rapid runoff (see Peters, 1985). In contrast, Panther Lake basin is a ground-water dominated-system in which a significant portion of incident precipitation moves through the ground water system as baseflow (Figure 5). The thick and relatively permeable upper-till deposits in Panther provide for extensive ground water storage and produce a uniform baseflow.

The longer residence time of water in the ground water system of Panther Lake allows the rate-limited neutralization reactions such as primary mineral weathering to proceed further toward equilibrium. Sufficient acid neutralization takes place to keep the waters of Panther Lake near a pH of 7 for most of the year. Only during the annual springmelt, when a large portion of water enters the lake through overland flow and shallow interflow, does the pH of Panther Lake surface water (upper 1 m) drop to approximately 5.

The water of Woods Lake is acidic because most incident acidic precipitation reaches the lake by surface and shallow interflow. The water has a short residence time in the thin soils and glacial till comprising the surficial deposits of the basin. Consequently, neutralization of acidic waters is incomplete. Woods Lake maintains a pH that, for the most part, ranges between 4.4 to 4.9 yr round.

3.2. CHEMICAL WEATHERING

Whether the influx of atmospheric acids has altered the rates of chemical weathering in drainage basins of the northeastern United States or not is currently being addressed. Measurements made during the course of the ILWAS project enabled us to determine the present rate of chemical weathering in Woods and Panther lake-watersheds and to estimate the long-term average weathering rate since the glaciers retreated from the area approximately 14 000 yr ago. The results of our initial analysis (April et al., 1985) suggest that in the Panther Lake drainage basin, chemical weathering has increased appreciably

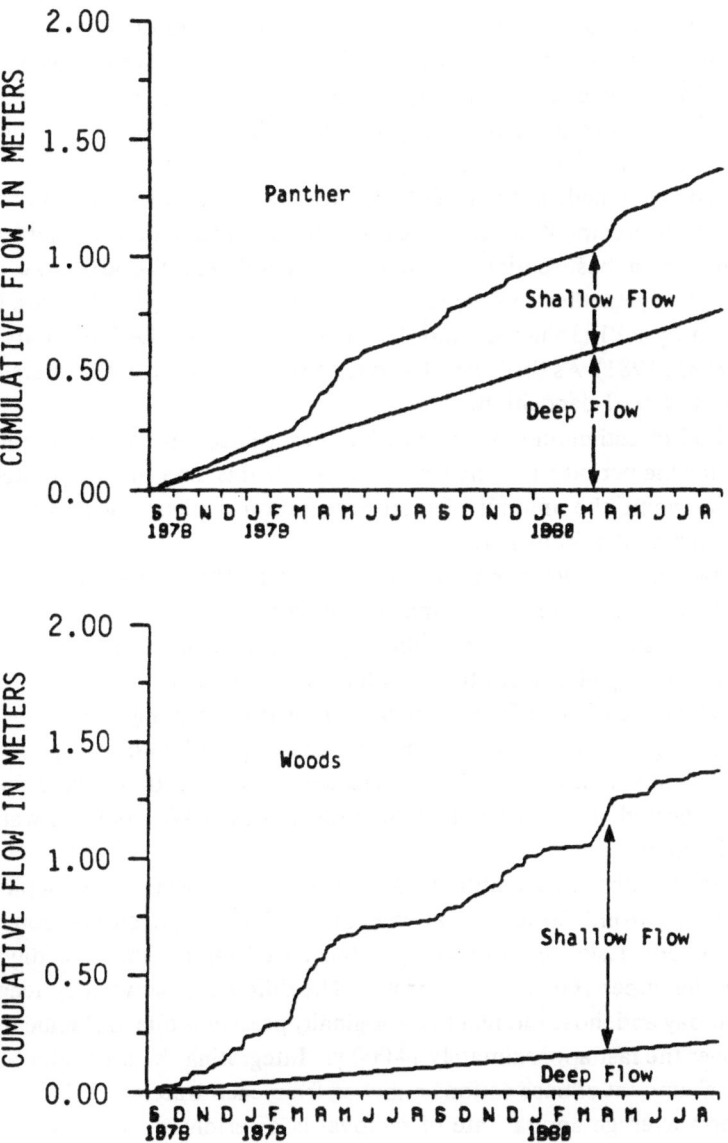

Fig. 5.   Cumulative lake inflows from Woods and Panther watersheds. Calculations were made using the
hydrologic module of the ILWAS model.

compared with the average long-term rate. In the Woods Lake drainage basin, chemical
weathering rates appear not to have increased at all, rather rates may have declined since
deglaciation.

The current rate of chemical weathering in a watershed may be expressed as the net
rate of release of base cations from the basin per unit area. This rate includes cations
released by primary silicate weathering, as well as those released from cation exchange

reactions. Johnson *et al.* (1977) found that the current rate of base cation release to streams (with a mean pH of about 6.7) in northern New England averages 2200 eq ha^{-1} yr^{-1} whereas for streams (pH $\sim$ 7) in North America the average is 3800 eq ha^{-1} yr^{-1}. Likens *et al.* (1977) reported a base cation release rate of 1000 to 1200 eq ha^{-1} yr^{-1} for Hubbard Brook (pH $\sim$ 5) in New Hampshire.

For the two watersheds in the ILWAS project, the current base cation release rates, determined from precipitation and stream water chemistry data by subtracting base cation inputs from base cation outputs, are as follows: Woods Lake (pH $\sim$ 4.7), 198 eq ha^{-1} yr^{-1}, Panther Lake (pH $\sim$ 7), 1679 eq ha^{-1} yr^{-1}. Values are based upon data collected by RPI (Johannes and Altwicker, 1981) and the University of Virginia (Galloway *et al.*, 1981). As discussed by Johannes *et al.* (1985), both watersheds receive similar inputs of acid deposition.

One method of estimating an average long-term base cation release rate for these watersheds for the period since the last glaciation of this area (approximately 14 000 yr ago) is to determine the amount of total base cation depletion within the soils. This approach assumes the following:

(1) Immediately following the retreat of the last glacier, no vertical gradation in elemental abundance existed in the mineral matter.

(2) The storage of base cations in the vegetation is insignificant.

(3) All cations depleted from the soils have left the basin.

(4) There has been insignificant soil erosion at the sampling sites.

(5) Weathering below a depth of 1 m is small compared with the upper 1 m of soil.

To support this analysis, bulk soil elemental compositions were measured in 87 samples collected at various soil depths from 9 sites in Woods Lake watershed and 11 sites in Panther.

Based upon the above assumptions, Figure 6 shows typical cation depletion curves for a soil profile from Panther Lake watershed. (Depletion curves constructed for profiles in Woods Lake are similar.) All the base cations show readily detectable depletion in the upper 100 cm of the profile. The difference between concentrations at the surface today and those thought to be originally present is due to chemical weathering occurring over the last approximately 14 000 yr. Integrating the area under the curves* gives the total amount of base cations removed from each watershed since deglaciation. The long-term average annual rate of removal is determined by dividing base cation depletion by the total number of years since deglaciation (Table V). Immediately after glacial retreat, the fresh, fine-grained detritus may have weathered at a higher rate than the long-term average. With time and the development of the soil, the weathering rate most likely decreased.

When current rates of cation release, as determined by input/output ion balances, are compared with long-term rates calculated from trends in soil bulk chemistry, we observe

---

* The long-term rate of cation release is calculated as follows: Integrated area (cm wt. % oxide) x % cation in oxide (corrected using Ti as a conservative constituent) x bulk density (g cm^{-3}) = g cation removed cm^{-2} of watershed covered by soil. Multiply by area of watershed covered by soil then divide by 14 000 yr to get rate of release for each cation species. Add cation rates in equivalents to get total release rate.

Panther Lake Watershed (Site 103)

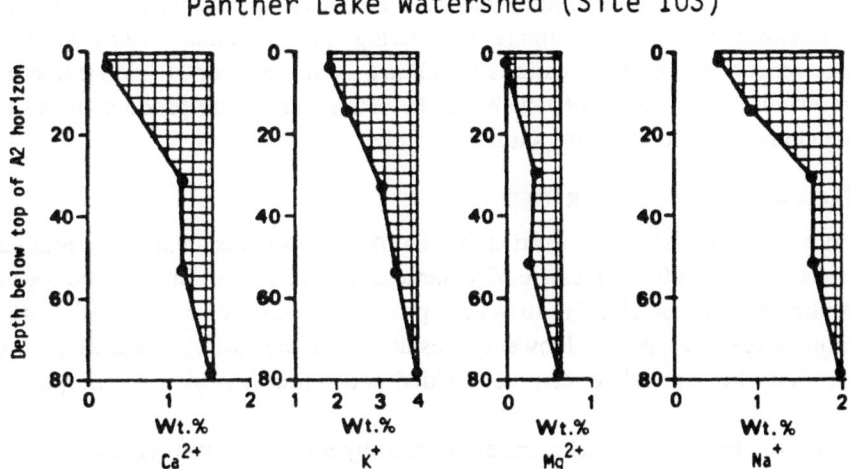

Fig. 6. Examples of base cation depletion curves for a soil profile in Panther Lake watershed.

TABLE V

Average annual long-term cationic release rates (eq ha^{-1} yr^{-1} of watershed not including outcrop areas.)

	Woods	Panther
Ca^{2+}	178 (142)[a]	100 (72)
Mg^{2+}	122 (84)	84 (42)
Na$^+$	189 (108)	165 (153)
K$^+$	129 (80)	150 (177)
Total	618 ($\geq$ 248)	499 ($\geq$ 248)

[a] Numbers in parentheses are the standard deviations. The limits on the standard deviations of the totals assume either positive or zero.covariances of the concentrations.

that the current cation release rate in Panther watershed is approximately 3 times greater than the long-term average. Application of the t-test shows that these rates are significantly different from one another. In contrast, the current cation release rate in Woods watershed is less than the long-term average. If the cause of the currently higher cation release rate in Panther Lake watershed is acid deposition, we suggest that the different response of the two watersheds to similar acid loadings is attributable to watershed hydrology and geology. The overall effect is that Woods Lake may have become acidic while Panther Lake remains neutral. In Panther, weathering rates appear to have adjusted proportionately to an increased influx of acidity. Weathering reactions and the production of alkalinity in Woods watershed, however, are not sufficient to fully neutralize current acidity loadings. We suggest that in Woods, the mineral soil layer does not have sufficient flow capacity to contact and neutralize all of the incident precipitation. The result is that some of the atmospheric acidity enters the lake.

Finally, it should also be pointed out that present cation release rates include both cations released from primary mineral weathering and base cations released by cation exchange reactions. The base cations which are exchanged for H^+ ion come from a finite reservoir, the capacity of which can be estimated by the product of the cation exchange capacity and the base saturation index.

3.3. Sagamore Lake Watershed

Sagamore Lake watershed, the third in the ILWAS project, has an area of about 42 km^2 and a relief of over 500 m. Because of its large size, difficult terrain and heterogeneous geomorphology (all contributing to access problems), seismic and bedrock data from its remote areas are sparse. However, results from the geologic mapping program indicate that the watershed can be divided into three distinct geomorphic areas (Figure 7):

(1) the Lost Brook subcatchment dominated by swampy lowlands and till deposits greater than 3 m in thickness;

(2) the East Inlet subcatchment which is an area underlain by thin till and bedrock; and

(3) the outwash plain adjacent to Sagamore Lake which is composed mainly of stratified sand.

Sagamore Lake watershed, therefore, cannot be classified simply as a thin till (Woods-like) system or a thick till (Panther-like) system. Rather, the watershed has characteristics of both Woods and Panther Lake systems. It is not surprising then that the response of Sagamore Lake to acidic deposition appears to be intermediate to those of Woods and Panther Lakes.

4. Summary

The soils and till in Woods and Panther lake watersheds contain quartz, potassium feldspar, plagioclase feldspar and hornblende. Accessory minerals included magnetite, ilmenite, hypersthene, garnet, tourmaline, epidote and zircon. Vermiculite was the predominant clay mineral in the soils and unconsolidated glacial till deposits. Vermiculite in the soil horizons contained Al interlayers that may be a potential source or sink of Al for the ground water. The high extractable Fe content in these soils (up to 6 weight-%) resulted from the prior chemical dissolution of hornblende.

The differences in the lake water chemical characteristics between Wood and Panther lakes result from differences in surficial geology. Although both basins are underlain by granitic bedrock and receive similar inputs of atmospheric acidity, Panther watershed is dominated by thick, relatively permeable glacial till whereas Woods watershed is composed of thin till with an aeolian mantle and has lower permeabilities. These differences affect the flow paths by which water reaches each lake. In Woods Lake watershed, most water moves to the lake as overland flow or shallow interflow, and little acid neutralization occurs. In Panther Lake watershed, a significant proportion of water moves through the deep glacial till. Within the ground-water system, weathering and cation exchange reactions consume hydrogen ions and produce alkalinity.

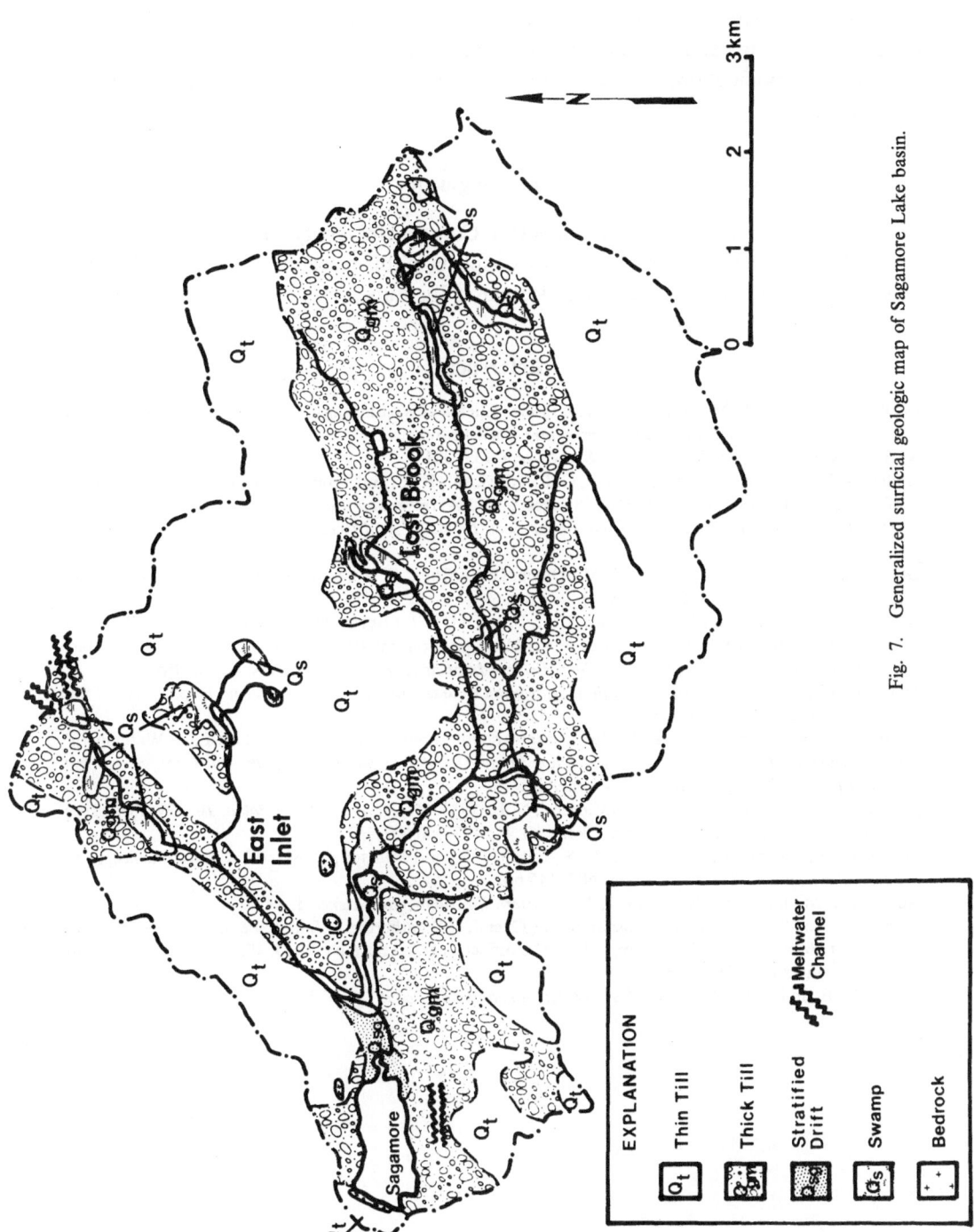

Fig. 7.   Generalized surficial geologic map of Sagamore Lake basin.

   The study of the ILWAS basins has shown that surficial geologic materials, even in the absence of carbonate minerals and in areas of sensitive bedrock, can provide adequate buffering capacity to offset lake acidification. To determine the susceptibility of a lake to acidification, one must characterize the surficial materials in its tributary watershed.

Acknowledgments

We would like to thank the many students at Colgate University and Smith College who assisted with both the field and laboratory aspects of this research. Funding for ILWAS was provided by the Electric Power Research Institute.

References

April, R. A., Newton, R. M., and Truettner, L.: 1985, 'Chemical Weathering in Two Adirondack Watersheds: Past and Present-day Rates', *Geol. Soc. Amer. Bull.* (in review).

April, R. A. and Newton, R. M.: 1983, *Soil Science* **135**, 301.

Galloway, J. N., Schofield, C. L., Hendrey, G. R., Altwicker, E. R., and Troutman, D. E.: 1981, 'An Analysis of Lake Acidification Using Annual Budgets', in *The Integrated Lake-Watershed Acidification Study (ILWAS): Contributions to the Int. Conf. on the Ecol. Impact of Acid Dep. EPRI Interim Report*, Palo Alto, CA.

Gherini, S. A., Mok, L., Hudson, R. J. M., Davis, G. F., Chen, C. W., and Goldstein, R. A.: 1985, *Water, Air, and Soil Pollut.* **26**, 425 (this issue).

Hendrey, G. R., Galloway, J. N., Norton, S. A., Schofield, C. L., Burns, D. A., and Schaffer, P. W.: 1980, 'Sensitivity of the Eastern United States to Acid Precipitation Impacts on Surface Waters', in D. Drablos and A. T. Tollan (eds.), *Proc. of Int. Conf. Ecolog. Impacts Acid Precip.* Norway, SNSF Project.

Jackson, M. L.: 1974, *Soil Chemical Analysis: Advanced Course*, University of Wisconsin, Department of Soil Science, Madison, WI. 895 pp.

Johannes, A. H. and Altwicker, E. R.: 1981, 'Atmospheric Inputs to Three Adirondack Lake-Watersheds', in *The Integrated Lake-Watershed Acidification Study (ILWAS): Contributions to the Int. Conf. on the Ecol. Impact of Acid Dep. EPRI Interim Report*, Palo Alto, CA.

Johannes, A. H., Altwicker, E. R., and Clesceri, N. L.: 1985, *Water, Air, and Soil Pollut.* **26**, 339 (this issue).

Johnson, N. M., Reynolds, R. C., and Likens, G. E.: 1977, *Science* **177**, 514.

Likens, G. E., Bormann, H. R., Pierce, R. S., Eaton, J. S., and Johnson, N. M.: 1977, *Biogeochemistry of a Forested Ecosystem*, Springer Verlag, NY. 146 pp.

Newton, R. M. and April, R. H.: 1982, *Northeastern Environmental Science* **1**, 143.

Parnell, R. A.: 1981, 'Aluminum Migration and Chemical Weathering in Subalpine and Alpine Soils and Tills. Mt. Moosilauke, New Hampshire: The Effects of Acid Rain', Ph.D. dissertation, Dartmouth College, Hanover, NH. 285 pp.

Peters, N. E. and Murdock, P.S.: 1985, *Water, Air, and Soil Pollut.* **26**, 387 (this issue).

HYDROGEOLOGIC COMPARISON OF AN ACIDIC-LAKE BASIN WITH A NEUTRAL-LAKE BASIN IN THE WEST-CENTRAL ADIRONDACK MOUNTAINS, NEW YORK

NORMAN E. PETERS

U.S. Geological Survey, Water Resources Division, Atlanta, GA 30360, U.S.A.

and

PETER S. MURDOCH

U.S. Geological Survey, Water Resources Division, Albany, NY 12201, U.S.A.

(Received November 1, 1984; revised June 18, 1985)

Abstract. Two small headwater lake basins that receive similar amounts of acidic atmospheric deposition have significantly different lake outflow pH values; pH at Panther Lake (neutral) ranges from about 4.7 to 7; that at Woods Lake (acidic) ranges from about 4.3 to 5. A hydrologic analysis, which included monthly water budgets, hydrograph analysis, examination of flow duration and runoff recession curves, calculation of ground-water storage, and an analysis of lateral flow capacity of the soil, indicates that differences in lakewater pH can be attributed to differences in the ground-water contribution to the lakes. A larger percentage of the water discharged from the neutral lake is derived from ground water than that from the acidic lake. Ground water has a higher pH resulting from a sufficiently long residence time for neutralizing chemical reactions to occur with the till. The difference in ground-water contribution is attributed to a more extensive distribution of thick till (<3 m) in the neutral-lake basin than in the acidic-lake basin; average thickness of till in the neutral-lake basin is 24 m whereas that in the other is 2.3 m. During the snowmelt period, as much as three months of accumulated precipitation may be released within two weeks causing the lateral flow capacity of the deeper mineral soil to be exceeded in the neutral-lake basin. This excess water moves over and through the shallow acidic soil horizons and causes the lakewater pH to decrease during snowmelt.

1. Introduction

Dilute concentrations of strong acids (sulfuric and nitric) in atmospheric precipitation have been reported in the northeastern United States (Cogbill and Likens, 1974; Cogbill, 1976) and southeastern Canada (Dillon *et al.*, 1978). The effect that this strong acidity has had on the composition of surface waters is largely unknown. A synoptic survey of lakes in the Adirondack Mountain region of northeastern New York (Figure 1) by Schofield (1976), however, found that 50% of the surveyed lakes at high altitudes (above 600 m) had a pH of less than 5 and that 90% were devoid of fish. In 1977, the U.S. Geological Survey, in cooperation with the University of Virginia and the Electric Power Research Institute, began a study known as the Integrated Lake-Watershed Acidification Study (ILWAS) to investigate why three lakes in the west-central Adirondack Mountains that receive similar amounts of acidic atmospheric deposition respond differently in neutralizing this acidity.

Water, Air, and Soil Pollution **26** (1985) 387–402.   0049–6979/85.15
© 1985 *by D. Reidel Publishing Company.*

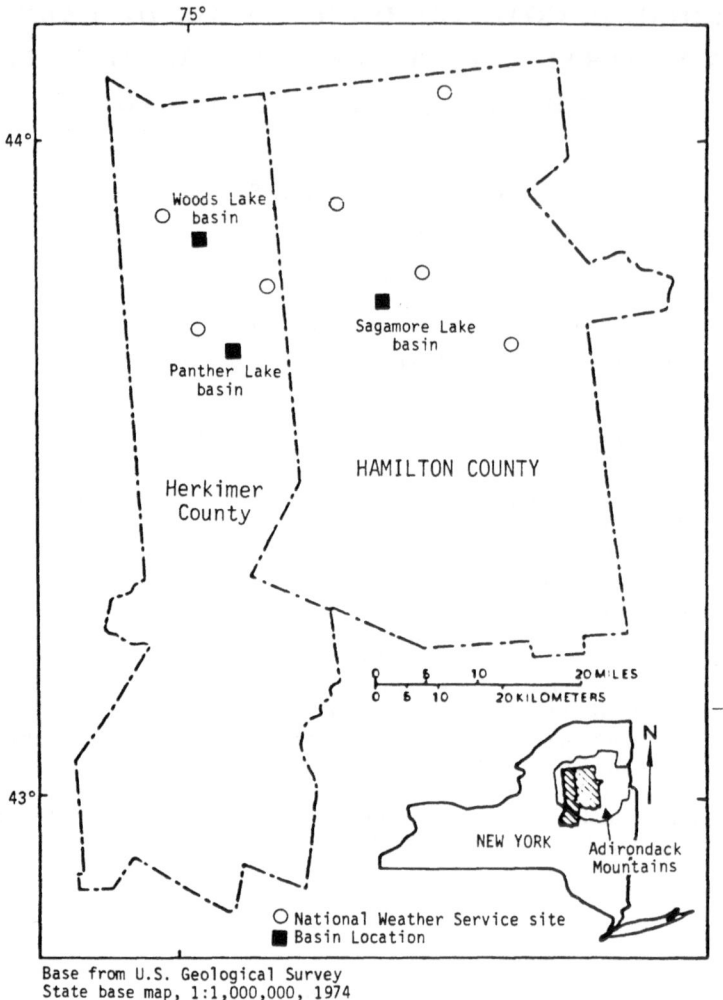

Fig. 1.   Location of Woods Lake (acidic) and Panther Lake (neutral), Herkimer County, NY.

    Analyses of data gathered in the first phase of ILWAS during 1977–80 suggest that
acid neutralization occurs within the watershed and is a function of the quantity and
rate of flow of water through surficial geologic materials (Troutman and Peters, 1982);
till is the primary surficial geologic material. To test the hypotheses generated from
research in the first phase of ILWAS, a second phase was begun in 1980 to further define
and compare the geohydrologic characteristics of two of the three ILWAS lake basins
(Figure 1). The two basins – Woods Lake (acidic) and Panther Lake (neutral) – have
similar physical dimensions and are underlain by relatively weathering-resistant granitic
gneiss, although they differ in lake water and lake-outflow pH. The pH of Panther Lake
ranges from about 4.7 to 7, and that of Woods Lake ranges from about 4.3 to 5.

This report summarizes the results of the hydrologic analysis of the two lake basins in the second phase of ILWAS from January 1980 through December 1981. Emphasis is placed on the flowpaths along which water moves through the watershed, because it was hypothesized that water that moves over or through the acidic upper soil horizons would be more acidic than water that moves through the till, where chemical reactions resulting in acid neutralization and alkalinity production can occur.

TABLE I

Basin and hydrologic characteristics of neutral and acidic lakes

	Panther Lake (neutral)	Woods Lake (acidic)
Basin area	1.24 km²	2.07 km²
Watershed area	1.06 km²	1.84 km²
Lake surface elevation above sea level	557 m	606 m
Lake area/basin area ratio	14%	11%
Relief	170 m	122 m
Amount of bedrock outcrop in basin*	10%	25–30%
Amount of bedrock and thin till in basin	33%	79%
Average till thickness	24.5 m	2.3 m
Lake volume, m³	7.08×10^5 m³	8.13×10^5 m³
Duration of ice cover on lake	120 d	120 d
Mean water detention time in lake	8.5 mo	7.3 mo
Mean water residence time in lake**	8–12 yr	0.7 to 1 yr
Average precipitation (1980–1981)	117 cm yr^{-1}	123 cm yr^{-1}
Average runoff (1980–1981)	72 cm yr^{-1}	76 cm yr^{-1}
Average evapotranspiration (1980–1981)	38%	38%

* Soil less than 20 cm in depth.

** Residence time $= \dfrac{\text{lake volume plus soil moisture volume}}{\text{average yearly precipitation}}$.

2. Hydrologic Characteristics

Both lake basins receive approximately the same amount of precipitation per unit area; Woods Lake (acidic) recieves 1.25 m yr^{-1}, and Panther Lake (neutral) receives 1.20 m yr^{-1}, based on 4 yr of record (Johannes et al., 1985). Most physical basin characteristics, including area, relief, and lake altitude are also similar, as shown in Table I. Both basins are nearly 100% forested, although they differ slightly in percentage of coniferous and deciduous trees (Cronan, 1985).

The two basins differ considerably in subsurface characteristics, however, particularly in the bedrock-surface configuration and the thickness of the till (Figure 2). The acidic-lake basin has a rough, 'washboard like' bedrock surface covered mostly by thin till less than 3 m thick; thicker till covers only 5% of the basin, and the average thickness for the basin is 2.3 m. The neutral-lake basin has a smooth, 'saucer like' bedrock surface, and till is thick, except on the upper slopes of Panther Mountain at the west side of the basin; average thickness for the basin is 24 m (April and Newton, 1985). A mantle of eolian silt overlies the till in part of the acidic-lake basin but not in the neutral-lake basin. Both basins are underlain by granitic gneiss and are believed to lose negligible amounts of water by underflow.

3. Water Budgets

For local flow systems with little interbasin transfer of water, precipitation must either be lost as evapotranspiration or flow through the watershed into the lake and then through the lake outlet. On a short-term basis, seasonal changes in water storage cannot be neglected, but over 2 yr, annual net change in storage is probably negligible. A water budget, computed from 2 yr of field observations and assuming no net change in storage, is shown below.

Average annual inflow/discharge 1980–81	Lake Basin	
	Woods (acidic)	Panther (neutral)
Precipitation, cm*	123	117
Outflow, cm*	76	72
Evapotranspiration (precipitation minus outflow)		
cm	47	45
% of precipitation	38	38

* Values from Johannes et al., 1985.

Budgets designed for monthly analysis should include changes in storage. The following equation can be used:

$$P = R + E + \Delta S_S + \Delta S_L + \Delta S_W \tag{1}$$

where

P = precipitation
R = runoff (discharge from the lake-watershed system)
E = evapotranspiration
ΔS_S = change in snowpack storage
ΔS_L = change in lake storage, and
ΔS_W = change in watershed storage.

From the field data and the evapotranspiration predicted by the vapor-pressure method, Equation (1) was solved for the change in watershed storage, ΔS_W. The results of these calculations for the two basins are shown in Tables II and III. Watershed storage includes ground water, soil moisture, and water held in surface depressions.

TABLE II

Monthly water balance (cm) for Panther Lake (neutral) basin [Values are in cm of water. P is precipitation measured by Rensselaer Polytechnic Institute (RPI), with no adjustments for missing values; ΔS_S is change in snowpack storage measured by RPI and USGS; R is runoff measured by USGS; ET is evapotranspiration calculated by vapor-pressure method; ΔS_L is change in lake storage measured by USGS and University of Virginia; and ΔS_W is the change in watershed storage derived using Equation (1).]

	P	ΔS_S	R	ET	ΔS_L	ΔS_W
1980						
January	8.25	+ 3.56	5.21	0.00	0.00	− 0.51
February	3.07	+ 3.05	3.07	0.00	− 0.23	− 2.82
March	12.85	− 2.29	6.86	0.00	+ 0.96	+ 7.31
April	10.18	− 4.32	12.83	3.28	− 0.89	− 0.71
May	5.66	−	4.01	6.37	− 0.66	− 4.06
June	11.68	−	6.43	7.67	+ 0.23	− 2.64
July	15.57	−	3.96	10.24	+ 0.23	+ 1.14
August	5.43	−	2.34	8.89	− 0.79	− 5.00
September	11.18	−	3.96	7.42	+ 0.13	− 0.33
October	12.45	−	3.73	2.34	+ 1.24	+ 5.13
November	11.33	−	7.64	0.00	− 0.15	+ 3.83
December	8.38	+ 7.62	6.25	0.00	− 0.66	− 4.83
Yearly	116.03	+ 7.62	66.29	46.20	− 0.58	− 3.48
1981						
January	2.41	+ 1.98	3.03	0.00	− 0.71	− 1.89
February	15.39	− 6.10	11.35	0.00	+ 2.08	+ 8.06
March	7.16	− 2.54	5.99	0.00	+ 0.61	+ 3.10
April	9.91	− 0.96	12.04	3.33	− 1.60	− 2.90
May	5.23	−	5.03	6.30	− 0.71	− 5.39
June	8.53	−	3.05	8.69	− 0.00	− 3.21
July	11.86	−	2.97	10.39	+ 0.08	− 1.58
August	13.44	−	4.88	7.95	+ 0.08	+ 0.53
September	18.11	−	7.03	4.52	+ 0.84	+ 5.72
October	16.00	−	10.34	2.31	+ 0.79	+ 2.56
November	5.71	−	7.19	0.00	− 1.02	− 0.46
December	4.16	+ 4.16	4.44	0.00	− 0.05	− 4.39
Yearly	117.91	− 3.46	77.34	43.49	0.39	0.15
Two years	233.94	+ 4.16	143.63	89.69	− 0.19	− 3.3
Percentage	−	1.8	61.4	38.3	negligible	1.4

Precipitation, runoff, change in snowpack storage, and change in lake storage were obtained from measured values for the two lake basins: precipitation quantity was monitored using an onsite weighing-bucket rain gage (Johannes et al., 1981); runoff was calculated from water-stage data derived from a water-stage recorder and a relationship between stage and measured discharge (Peters et al., 1985); depth and water equivalent of the snowpack were determined using an Adirondack snow coring device (Johannes et al., 1981; Peters et al., 1985); and lake water levels were derived from staff-gage readings (Peters et al., 1985).

TABLE III

Monthly water balance (cm) for Woods Lake (acidic) basin [Values are in cm of water. P is precipitation measured by Rensselaer Polytechnic Institute (RPI), with no adjustments for missing values; ΔS_S is change in snowpack storage measured by RPI and USGS; R is runoff measured by USGS; ET is evapotranspiration calculated by vapor-pressure method; ΔS_L is change in lake storage measured by USGS and University of Virginia; and ΔS_W is the change in watershed storage derived using Equation (1).]

	P	ΔS_S	R	ET	ΔS_L	ΔS_W
1980						
January	8.13	+ 3.81	4.16	0.00	− 2.23	+ 0.38
February	3.30	+ 2.72	0.99	0.00	− 0.41	+ 0.00
March	11.45	− 3.12	7.87	0.00	+ 1.12	+ 5.59
April	10.72	− 3.40	13.87	3.58	− 0.48	− 2.84
May	7.06	−	2.44	6.78	+ 0.23	− 2.39
June	9.55	−	3.55	8.48	− 0.58	− 1.90
July	16.03	−	2.97	10.77	+ 0.41	+ 1.88
August	6.96	−	1.29	9.55	+ 0.20	− 4.09
September	11.30	−	2.34	5.54	+ 0.20	+ 3.22
October	13.11	−	6.43	2.11	+ 0.10	+ 4.47
November	12.85	−	10.84	0.00	+ 0.08	+ 1.93
December	10.49	+ 7.97	6.88	0.00	− 0.20	− 4.1
Yearly	120.95	+ 7.97	63.65	46.81	− 0.43	+ 2.08
1981	2.59	+ 2.59	1.42	0.00	− 0.02	− 1.40
January	15.72	− 7.24	19.15	0.00	+ 0.13	+ 3.68
February	7.52	− 2.54	8.03	0.00	+ 1.29	+ 0.74
March	8.58	− 0.79	10.08	3.43	− 1.29	− 2.85
April	3.96	−	2.34	6.15	+ 0.02	− 4.55
May	11.76	−	4.52	10.16	− 0.33	− 2.59
June	14.43	−	4.52	9.60	− 0.02	+ 0.33
July	13.00	−	9.07	7.39	− 1.32	− 2.13
August	16.69	−	6.63	4.98	+ 1.22	+ 3.86
September	18.16	−	13.99	2.51	+ 0.08	+ 1.57
October	7.44	−	4.98	0.00	− 0.20	+ 2.67
November	4.62	+ 4.62	3.02	0.00	− 0.18	− 2.84
December						
Yearly	124.47	− 3.35	87.76	43.49	− 0.63	− 3.51
Two years	245.42	+ 4.62	151.41	91.03	− 0.20	− 1.43
Percentage	−	1.9	61.7	37.0	negligible	0.6

To select a method for computing evapotranspiration, monthly evapotranspiration calculated by each of four different empirical formulas was summed to give annual values, these were compared with those derived from the annual water budgets. The annual evapotranspiration value that was the most similar to that derived from the annual water budget was the value calculated by the saturation vapor-pressure method of Hamon (1961), (Murdoch *et al.*, 1985). This estimate of evapotranspiration contains a undeterminable error that affects the accuracy of the residual change in watershed storage. However, since the basins have similar physical characteristics and were near each other (less than 30 km), differences in the actual evapotranspiration were believed

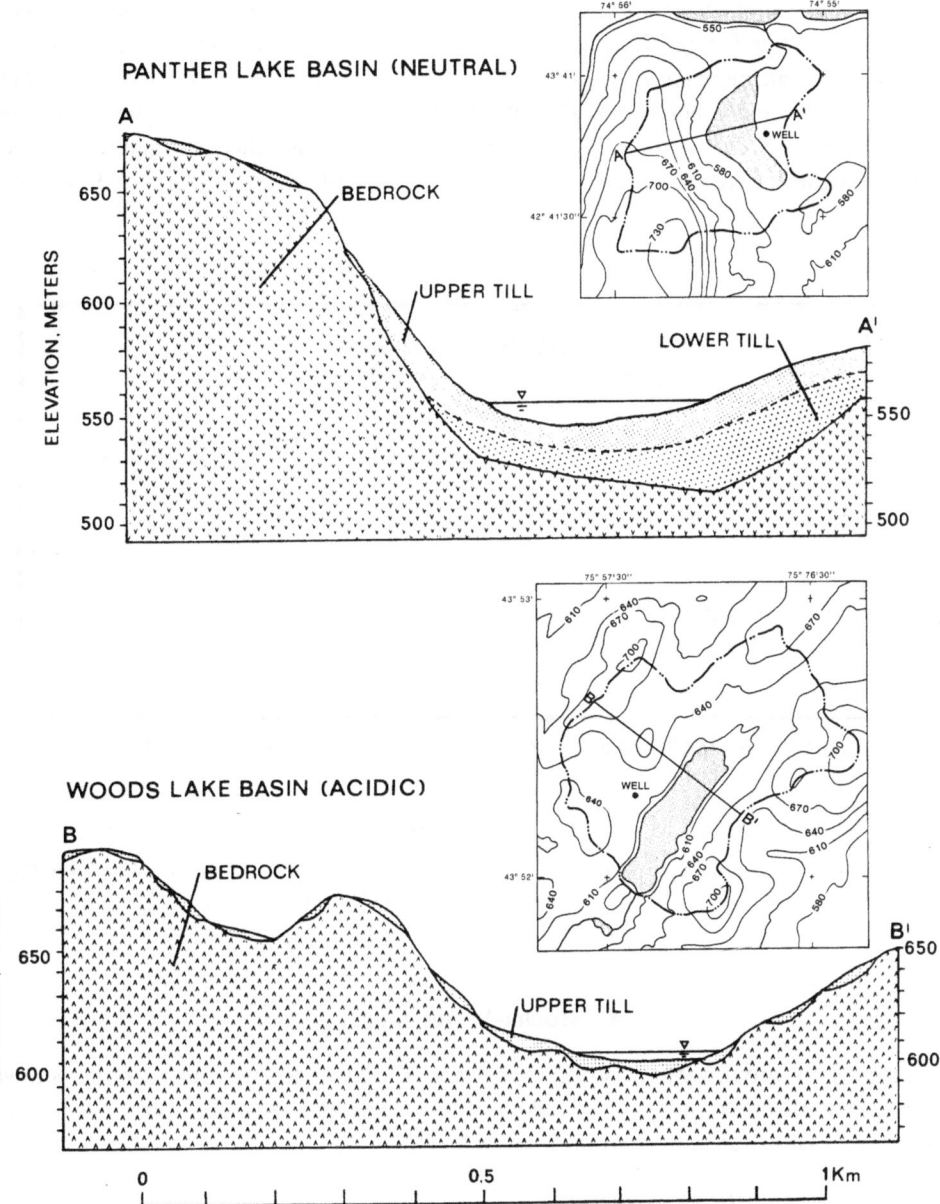

Fig. 2. Generalized geologic cross sections: (a) Panther Lake (neutral) basin; (b) Woods Lake (acidic) basin. (Locations are shown in Figure 1.)

to be small. Consequently, the calculated value for a month in one basin should have a deviation from the actual evapotranspiration that is comparable to that for the other basin.

Recharge of the ground-water system is indicated when ground-water levels increase (Figure 3). Recharge occurs primarily during the early spring and late fall, when water loss through evapotranspiration is small; recharge during the spring coincides with snowmelt, and that during fall with storms. The increase in ground-water levels generally coincides with an increase in watershed storage; compare the changes in watershed storage for 1980 in Tables II and III with the ground-water levels in Figure 3. During the spring and fall of 1980–81, the monthly changes in watershed storage were larger per unit area in the neutral-lake basin than in the acidic-lake basin (Tables II and III).

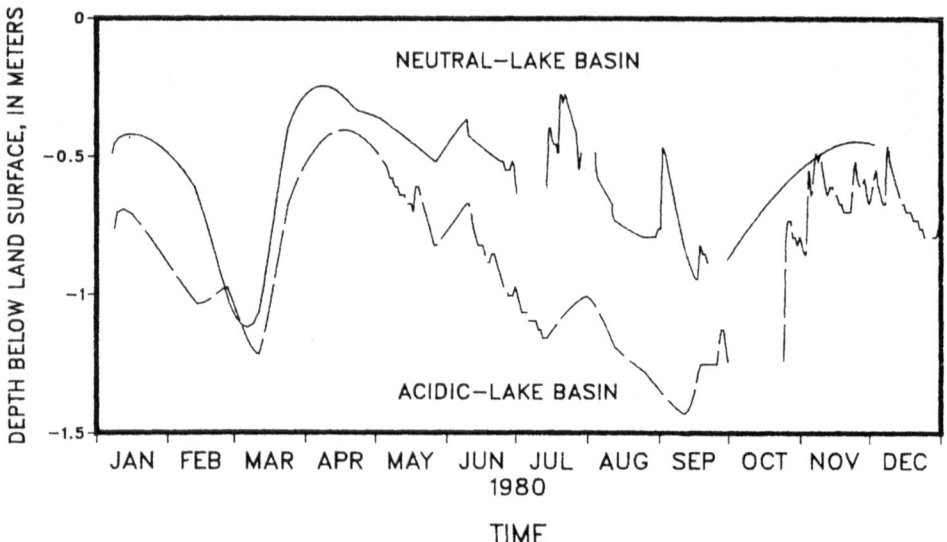

Fig. 3.   Comparison of ground water levels in thick till of neutral and acidic lake basins during 1981. (Well locations shown in insets, Figure 2.)

4. Ground-Water Storage

Differences in ground-water storage between the two basins can also be shown qualitatively through analysis of lake-outflow hydrographs, flow-duration curves, outflow-recession rates, and ground-water levels.

4.1. HYDROGRAPHIC ANALYSIS

Outflow hydrographs indicate that discharge rates at the acidic-lake outlet increase and decline rapidly in response to precipitation (Figure 4), which is characteristic of watersheds through which precipitation moves quickly as surface runoff or shallow interflow (lateral flow through upper soil horizons). Outflows from the neutral lake increase similarly, but do not decline as rapidly, which is indicative of hydrologic systems in which the precipitation percolates into deeper soil horizons. Watersheds of this type tend to have lower peak flows and higher base flows (Gray, 1970).

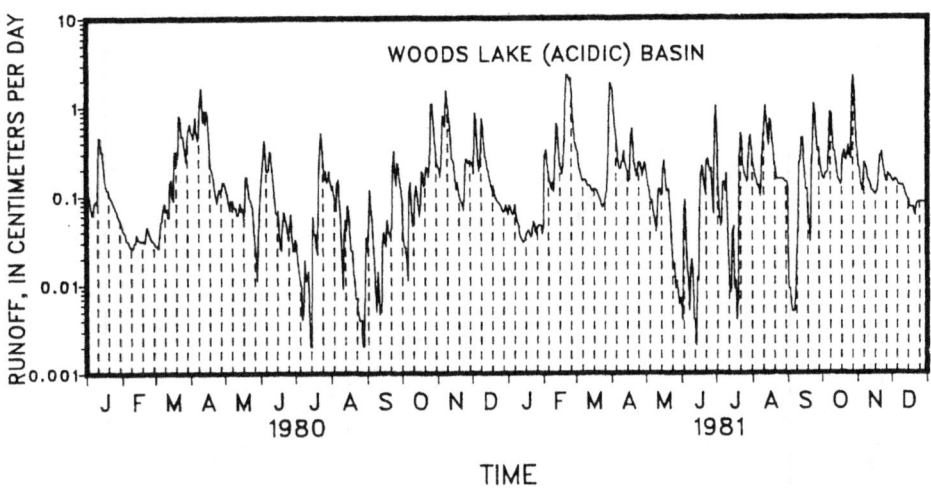

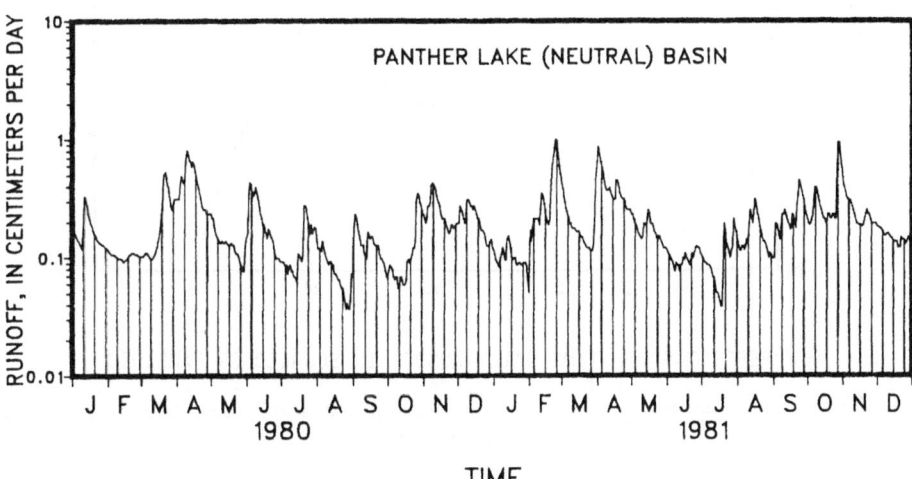

Fig. 4.   Hydrographs of daily runoff, January 1980 through December 1981: (a) Panther Lake (neutral) basin; (b) Woods Lake (acidic) basin.

4.2. FLOW-DURATION ANALYSIS

The differences in the proportion of base flow (deep ground-water flow) in the two basins can also be illustrated through flow-duration curves. Such curves show the percentage of time a specific discharge (basin outflow) is equaled or exceeded. For a basin that has highly variable discharge and a large proportion of surface runoff, the curve will have a steep slope. For a basin where stream or lake outflows are derived mostly from ground water, the curve will have a flatter slope. Storage of water in lakes and swamps will also tend to attenuate the flow and produce a flatter flow-duration curve (Searcy, 1959). This was not considered to be a factor affecting differences in the

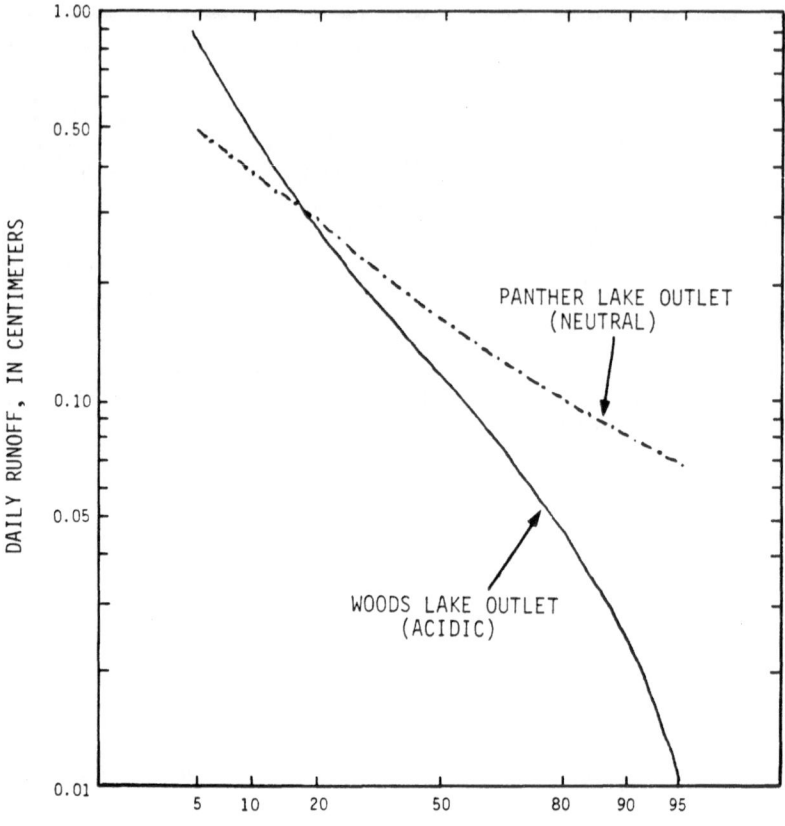

Fig. 5. Flow-duration curves for daily runoff from neutral and acidic lakes, January 1980 through
December 1981.

curves, however, because the percentage of lake area is similar between basins. Figure 5
presents such flow-duration curves for 1980 through 1981 at the two lake outlets. The
curve for the acidic lake has a steep slope, which suggests a predominance of surface
runoff and interflow and a smaller ground-water contribution. The curve for the
neutral-lake outlet has a flatter slope, which suggests a larger ground-water storage
capacity.

4.3. RECESSION-RATE ANALYSIS

The base flows and ground-water storage for the two basins can be calculated from
recession-rate curves (the rate at which flow decreases following a precipitation event)
through the method of Knisel (1963). Recession-rate curves for the lake outflows can
be evaluated only for the winter, when precipitation is in the form of snow, because
during the rest of the year rainstorms are too frequent – about once every three days

(Johannes *et al.*, 1981) – for a sufficiently long flow recession to occur. For the lake outflows, the maximum base flow normalized with respect to watershed area was $0.019 \, m^3 \, s^{-1} \, km^{-2}$ for the acidic-lake outlet and $0.039 \, m^3 \, s^{-1} \, km^{-2}$ for the neutral-lake outlet. During flow recessions, ground water is supplying most of the flow. The calculated sustained annual ground-water storage needed to supply the above base flows was 1.8 cm for the acidic lake and 6.9 cm for the neutral lake, again indicating more active ground-water storage in the neutral-lake basin.

4.4. GROUND-WATER LEVELS

Another technique for evaluating ground-water contributions to lakes is based on fluctuations in ground-water levels in the watersheds (Figure 3). The minimum volume of ground-water storage that is available to attenuate runoff from the two lakes can be estimated from the overall decline in water level during the summer of 1980 and the contributing thick till area. Note, however, that such an analysis is only as valid as the representativeness of the wells available for use. The data and assumptions listed below have been used to calculate the ground-water-storage capacity in the thick till of both lake basins:

(1) Ground-water declines occur in areas adjacent to the lake that contain thick till (greater than 3 m) – 53% of the neutral-lake watershed ($1.06 \, km^2$) and 10% of the acidic-lake watershed ($1.84 \, km^2$).

(2) The ground-water level dropped 0.7 m from mid-April to mid-September of 1980 in the neutral-lake watershed an 1 m during the same period in the acidic-lake watershed as represented by the ground-water decline in one well in thick till at each basin.

(3) Specific yield of the till (volume of water discharged per unit area under a unit drawdown) is 0.2 in both watersheds.

(4) Total basin area for the neutral-lake watershed is $1.24 \, km^2$; that of the acidic-lake basin is $2.07 \, km^2$.

During the summer, the overall decline in ground-water levels in the thick till of the neutral-lake basin was equivalent to 6 cm applied uniformly across the basin as opposed to only 2 cm for the acidic-lake basin. Note from Figure 3 that the decline was not continuous, and significant recharge occurred in June and late July, which indicates that much more water actually moved through these ground-water systems. These values, therefore, represent the minimum capacity of the ground-water systems to control and attenuate flows into the lakes.

5. Lateral Flow Capacity of the Saturated Till

Another approach for estimating how much water follows a deep flowpath compared with a shallow flowpath is based on the maximum lateral flow capacity of the till. For precipitation to flow through the till to the lake, the water must escape evapotranspiration and retention in upper soil horizons and reach the zone of saturation. At that point, the water begins to move laterally to the lake. The rate at which such water moves can be calculated from Darcy's Law:

$$Q_L = KAI, \qquad (2)$$

where

Q_L = lateral flow through the till,
K = saturated hydraulic conductivity of the till,
A = the flow cross-sectional area, and
I = hydraulic gradient.

The K value of till in the acidic-lake basin has been estimated to be 3.7×10^{-4} cm s^{-1} (April and Newton, 1985). The cross-sectional area A, as a first approximation, can be calculated from the perimeter of the lake, approximately 3000 m, and an average thickness of the till at that point, 1.5 m. The hydraulic gradient, again as a first approximation based on the slope of the land surface adjacent to the lake, is about 0.12. Entering these numbers into the equation above and converting units yields a volumetric flow capacity of 173 m^3 day^{-1}.

The K value of till in the neutral-lake basin has been estimated to be 5.3×10^{-3} cm s^{-1} (April and Newton, 1985). The cross sectional area A, was calculated from the length of shoreline, 2100 m, and the mean depth of the lake 3.9 m. The hydraulic gradient is about 0.10. From these data, Q_L for the neutral-lake basin is calculated to be 3,750 m^3 day^{-1}.

To translate the values of Q_L to equivalent precipitation, one simply divides Q_L by the surface of its tributary land. This area is 1.84×10^6 m^2 for the acidic-lake basin and about 1.06×10^6 m^2 for the other. Thus, the Q_L values in units of equivalent precipitation are 0.01 cm day^{-1} or 3.5 cm yr^{-1} for the acidic-lake basin and 0.35 cm day^{-1} or 129 cm yr^{-1} for the neutral-lake basin.

Because the annual precipitation in the area was approximately 120 cm yr^{-1} and the evapotranspiration rate about 45 cm yr^{-1}, 75 cm yr^{-1} of water flowed through the lakes. The Q_L value of 3.5 cm yr^{-1} for the acidic-lake basin indicates that only a small amount of the water that reached the lake followed a deep flowpath through the till. Because the vertical cross-sectional area through which the lateral flow can pass is small, most of the flow occurs as overland flow or is confined to lateral flow through the upper soil horizons.

For the neutral-lake basin, a Q_L value of 129 cm yr^{-1} indicates that lateral flow through the till can accommodate most precipitation falling on the watershed except during high intensity rainfall events or snowmelt periods. During the spring snowmelt period as much as three months of precipitation may be stored in the snowpack, and the accumulated water is released over a one- to two-week snowmelt period. The equivalent precipitation rate, including normal precipitation during the melt period, would be about 2.5 cm day^{-1} or 900 cm yr^{-1}. This analysis suggests that during the snowmelt period, the limit of Q_L for the neutral-lake basin is exceeded and part of the water seeks a shallower flowpath to the lake over or through the upper, more acidic soil horizons.

6. Factors that Affect Water Chemistry

Alkalinity and pH of the outflow at the acidic lake are much lower than those at the neutral lake. Average pH and alkalinity values for the outflow at the acidic lake were 4.7 and -10 µeq L^{-1}, respectively, whereas those at the neutral lake were 6.2 and 147 µeq L^{-1} (Schofield *et al.*, 1985). Large temporal changes do, though, occur in the pH and alkalinity of the neutral lake during snowmelt (Figure 6; Galloway *et al.*, 1983; Schofield *et al.*, 1985).

For both watersheds, the pH of precipitation ranges from about 3.5 to 4.5 and is altered somewhat by interaction with the forest canopy before it contacts the soil.

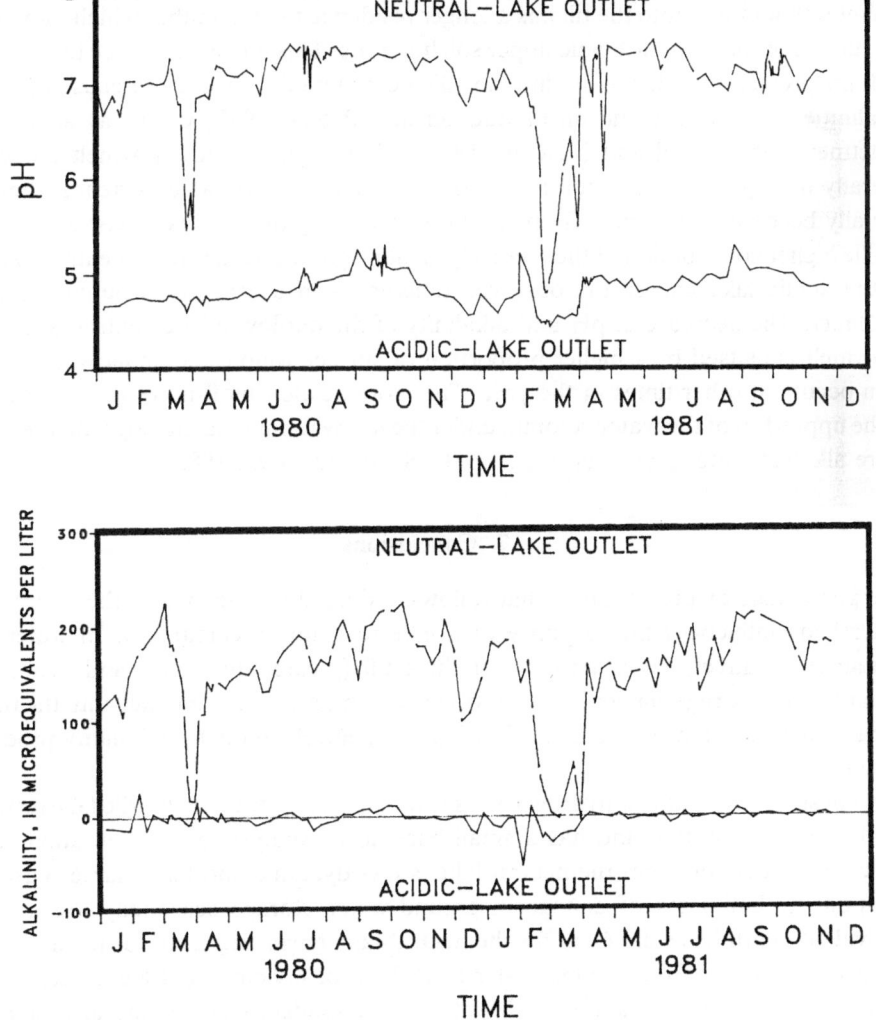

Fig. 6.   Chemical composition of the outflow at the neutral and acidic lakes, January 1980 through December 1981: (a) air-equilibrated pH; (b) alkalinity.

Throughfall under deciduous trees is more neutral than precipitation and that under coniferous trees is more acid (Johannes et al., 1985). As the water percolates into the deeper soil it becomes progressively less acidic with depth; soil water sampled from the boundary between the organic horizon and mineral horizon had pH's ranging from 3.5 to 3.9 and that from the lower B horizon, at a 50 cm depth, had an average pH of 4.5 (Cronan, 1985). Alkalinity and pH of groundwater sampled from wells screened at the water table were about 100 µeq L^{-1} and 6.0, respectively, whereas values for ground water sampled from a well screened 5 feet below the water table averaged 400 µeq L^{-1} and 7.5 (Peters et al., 1985). Furthermore, the chemical characteristics of soil and ground water from the acidic-lake basin were virtually indistinquishable from those from the neutral-lake basin.

Water that moves into the till has a longer residence time than that which moves over the surface or through the acidic upper soil horizons. This longer residence time together with the greater abundance of alumino-silicate minerals is thought to cause the higher alkalinities observed in the till (Troutman and Peters, 1982; Galloway et al., 1983; Truettner, 1984; April and Newton, 1985). If throughfall and snowmelt move only laterally through the upper soil horizons, particularly the organic horizons, they may actually become more acidic from reactions occurring there (Rosenqvist et al., 1980).

The higher contribution of the relatively alkaline ground water to the neutral lake than to the acidic lake causes the observed differences in average lake outflow pH and alkalinity. The decrease in pH and alkalinity of the outflow of the neutral lake during snowmelt is caused by a higher proportion of surface runoff and shallow lateral flow than occurs at other times of the year. The more acidic runoff flows through the lake in the upper 1 m of the water colomn, under the ice, with minimal mixing with the deeper more alkaline water (Hendrey et al., 1980; Schofield, et al., 1985).

7. Conclusions

The percentage of precipitation that follows a deep flowpath within the Adirondack watersheds influences the response of a lake to acidic precipitation. In watersheds containing relatively large amounts of thick till (greater than 3 m) and hence large ground-water storage capacities, more of the precipitation will percolate into the till and remain there for longer periods, allowing for neutralization by alkalinity-producing reactions.

Analysis of lake-outflow hydrographs shows that runoff from the acidic lake responds quickly to precipitation and has a small base flow. Analysis of flow duration curves indicates that runoff from the neutral lake is less dynamic and has a larger base-flow component. Estimates of change in ground-water storage seasonally, on a water-equivalent depth basis are 6 cm for the neutral-lake basin and 2 cm for the acidic-lake basin as calculated from ground-water level data, or 6.9 cm and 1.8 cm, respectively, based on recession rate analyses. Monthly water-budget calculations also show the neutral-lake basin to have higher ground-water recharge in fall and spring and greater declines in the winter than the acidic-lake basin. All the above indicate that the neutral-lake basin has a larger, more dynamic ground-water reservoir than the acidic-lake basin.

The lateral flow capacity of the till in both basins was estimated to determine whether this capacity is commonly exceeded by the rate at which water is applied to the soil. The maximum flow capacities in terms of applied precipitation were $3.5 \, \text{cm yr}^{-1}$ ($0.01 \, \text{cm day}^{-1}$) for the acidic-lake basin and $129 \, \text{cm yr}^{-1}$ ($0.35 \, \text{cm day}^{-1}$) for the neutral one. The actual net precipitation (the part not lost to evapotranspiration) in both basins was about $75 \, \text{cm yr}^{-1}$. Lesser amounts of precipitation entering the acidic-lake basin can pass through the till because of its smaller cross-sectional area and permeability than in the neutral-lake basin. Instead, more flow occurs over or through the shallow, acidic organic soil horizons.

The lateral-flow capacity of the neutral-lake basin would seem, at first glance, not to be exceeded; however, as much as three months of accumulated precipitation may melt within two weeks during the snowmelt period. This would be equivalent to precipitation at a rate of about $2.5 \, \text{cm day}^{-1}$ ($900 \, \text{cm yr}^{-1}$). Under these conditions, the lateral flow capacity of the mineral horizon in the neutral basin would most likely be exceeded. The resulting surface and shallow flow would be expected to be more acidic than the mineral horizon flow. The observed decrease in pH and alkalinity in the lake during snowmelt is consistent with this.

References

April, R. H. and Newton, R. M.: 1985, *Water, Air, and Soil Poll.* **26**, 373 (this issue).

Cogbill, C. V.: 1976, *Water, Air, and Soil Poll.* **6**, 407.

Cogbill, C. V. and Likens, G. E.: 1974, *Wat. Resour. Research* **10**, 1133.

Cronan, C. S.: 1985, *Water, Air, and Soil Poll.* **26**.

Dillon, P. J., Jeffries, D. S., Snyder, W., Reid, R., Yan, N. D., Evans, D., Moss, J., and Scheider, W. A.: 1978, *J. Fish. Res. Bd. Can.*, **35**, 809.

Galloway, J. N., Schofield, C. L., Peters, N. E., Hendry, G. R., and Altwicker, E. R.: 1983, *Can. J. Fish. Aquat. Sci.*, **40**, 799.

Gray, D. M.: 1970, *Handbook on the Principles of Hydrology*, Water Information Center, Inc. Port Washington, N.Y. 7.1.

Hamon, W. R.: 1961, *J. Hydrau. Div., ASCE* **87**, 207.

Hendrey, G. R., Galloway, J. N., and Schofield, C. L.: 1980, 'Temporal and Spatial Trends in the Chemistry of Acidified Lakes under Ice Cover', in D. Drablos and A. Tollan (eds.), *Ecological Impact of Acid Precipitation*, SNSF-Project, Norwegian Institute for Water Research, 266.

Johannes, A. H., Altwicker, E. R., and Clesceri, N. L.: 1985, *Water, Air, and Soil Poll.* **26**, 339 (this issue).

Johannes, A. H., Altwicker, E. R., and Clesceri, N. L.: 1981, 'Characterization of Acid Precipitation in the Adirondack Region, Electric Power Research Institute', Palo Alto, CA. EA-1826, 130.

Knisel, W. G.: 1963, *J. Geophys. Res.* **68**, 3649.

Murdoch, P. S., Peters, N. E., and Newton, R. M.: 1985, 'Hydrologic Analysis of Two Headwater Lake Basins of Differing Lake pH Precipitation in the West-Central Adirondack Mountains', New York, U.S. Geol. Survey Water Resources Investigations 84-4313 (in press).

Peters, N. E., Murdoch, P. S., and Dalton, F. N.: 1985, 'Hydrologic data of the Integrated Lake Watershed Acidification Study (ILWAS) in west-central Adirondack Mountains', New York from October 1977 through December 1981, U.S. Geological Survey Open-File Report 85-80 (in press).

Rosenqvist, I. Jorgensen, Th., P., and Rueslatten, H.: 1980, 'The Importance of Natural H^+ Production for Acidity in Soil and Water', in D. Drablos and A. Tollan (eds.), *Ecological Impact of Acid Precipitation*, SNSF-Project, Norwegian Institute for Water Research, 240.

Schofield, C. L.: 1976, 'Acidification of Achirondack Cokes by Atmospheric Precipitation: Extent and Magnitude of the Problem', Final Rep. D.J. Proj. F-28-4, NYS Dept., Env. Cons., 11 pp.

Schofield, C. L., Galloway, J. N., and Hendry, G. R.: 1985, *Water, Air, and Soil Poll.* **26**, 403 (this issue).

Searcy, J. K.: 1959, 'Flow-duration Curves – Manual of Hydrology: Part 2', Low-flow techniques, U.S. Geol. Survey Water-Supply Paper 1542-A, 33.

Troutman, D. E. and Peters, N. E.: 1982, 'Deposition and Transport of Heavy Metals in Three Lake Basins Affected by Acid Precipitation in the Adirondack Mountains, New York', in L. H. Keith (ed.), *Energy and Environmental Chemistry*, Vol. 2. Ann Arbor, Michigan, Ann Arbor Science Publishers, 33.

Truettner, L. E.: 1984, 'Mineral Weathering and Sources of Alkalinity in Two Adirondack Watersheds', M.S. Thesis, Smith College, Northampton, Mass., 147.

SURFACE WATER CHEMISTRY IN THE ILWAS BASINS

CARL L. SCHOFIELD

Department of Natural Resources, Fernow Hall, Cornell University, Ithaca, NY 14853, U.S.A.

JAMES N. GALLOWAY

Department of Environmental Sciences, Clark Hall, University of Virginia, Charlottesville, Va 22903, U.S.A.

and

GEORGE R. HENDRY

Terrestrial and Qquatic Ecology, Brookhaven National Laboratory, Upton, NY 11973, U.S.A.

(Received November 1, 1984; revised May 14, 1985)

Abstract. Alkalinity and pH differences observed between the three ILWAS lakes (Panther, Sagamore, and Woods lakes) are primarily a result of inherent watershed differences in base cation supply rates, relative to comparable strong acid input levels. The relatively high proportion of base rich ground water input to Panther Lake results in high pH and alkalinity (annual mean pH 6.2*, alkalinity 147 μeq L^{-1}). In contrast, shallow interflow with excess strong acid and high Al levels dominates the Woods Lake basin (annual mean pH 4.7*, alkalinity -10 μeq L^{-1}). Temporal acidification, observed in all three basins during snowmelt, occurs as a result of base cation dilution (particularly in Panther Lake) and increased strong acid anion levels. These marked changes in surface water chemistry are related to an upward shift in flow paths from ground water dominated base flow to shallow interflow during increased snowmelt discharge. Elevated nitrate and Al levels observed during these episodes suggest that HNO_3 from the snowpack and soil nitrification triggers acidification and Al mobilization. On an annual basis, Al export rates were less than 2% of total base cation output from the Panther basin and 23% of base cation output from the Woods basin. Woods Lake itself serves as a secondary sink for Al exported from the soils of the watershed, particularly during the summer months when increasing pH levels induce Al precipitation. Annually, an estimated 43% of the Al entering the lake is retained. Strong acid neutralization in the ILWAS basins appears to be a two-stage process, with initial Al mobilization in upper soil horizons followed by primary mineral dissolution and alkalinity production in deeper soil horizons. Separation of these processes in either time or space results in incomplete neutralization, acidification, and export of inorganic Al to surface waters.

1. Introduction

Extensive surveys of surface water quality were conducted in the Adirondacks during the 1970's to determine the regional extent of lake acidification and the chemical characteristics of acidified waters (Schofield, 1976; Pfeiffer and Festa, 1980). These surveys revealed a high proportion of acid lakes at higher elevations (50% of all lakes above 610 m were below pH 5), particularly in watersheds draining the western slopes of the Adirondack Mountains. Sulfate was found to be a major anionic constituent in all lakes sampled and high levels of Al were present in lakes with pH levels below 5. Subsequent field and laboratory studies demonstrated the significance of Al as a factor limiting fish survival in acid waters (Schofield and Trojnar, 1980; Baker and Schofield,

---

* pH-calculated from averaged H^+ concentrations, the latter measured as pH in the field.

1982). Although these survey and experimental studies were useful for inventory purposes and the establishment of fish management policy for acid waters on a regional basis, they did not provide sufficient information for critical evaluation of the processes contributing to lake acidification and Al mobilization. In particular, satisfactory explanations for the wide range in pH levels (~ 4 to 7) observed in lakes situated on similar bedrock formations and receiving comparable atmospheric acid deposition levels were not obtainable from the limited chemistry data available.

Lake or stream water chemistry observations represent an integrated measure of ecosystem modifications of water inputs from atmospheric precipitation. When these observations are evaluated in concert with other watershed information on flow pathways and chemical transformations affecting water transported through the drainage system, a basis is provided for the formulation and testing of hypotheses to explain surface water acidification. This was the approach taken in the ILWAS program. The three lakes selected for study are representative of the range in pH levels currently observed in Adirondack lakes (pH 4 to 7), yet all three are situated on simular bedrock and receive comparable atmospheric strong acid loadings. The primary objectives of the water quality phase of ILWAS were to determine the chemical nature and sources of acidity in the surface waters of the basins and provide sufficient chemical data for evaluation of the differences in water quality of the three lakes. Temporal acidification events observed during snowmelt periods provide additional, dynamic information for evaluation of hypotheses developed to explain inter-basin differences in water quality. These time variant water quality data were also utilized to evaluate potential sources and mechanisms of Al mobilization in the drainage basins. The amorphous $Al(OH)_3$ and exchangeable Al, which are abundantly present in the acid soils of the ILWAS basins, serve as sources for solution-phase Al which can be transported through soils and into streams and lakes. The Al is initially brought into solution by mineral and organic acids. The mineral acids are derived from atmospheric deposition and sources in the upper soil horizons. The organic acids are derived from the soil. Evaluation of the relative significance of these Al mobilization mechanisms and the importance of Al dissolution processes as an integral facet of the acidification process were of primary concern in this study. This paper summarizes the major findings of the surface water chemistry investigations described above.

2. Methods

Water samples were collected in acid washed polypropylene bottles from the outlets and major inlets of Woods, Sagamore, and Panther lakes on a weekly basis, with more frequent sampling (daily or bidaily) during spring runoff periods. Lake profile samples were collected with a peristaltic pump from six depths at monthly intervals and weekly during spring runoff. Water sampling was initiated in 1977 and continued through 1981.

Conductivity and pH determinations were performed on the day of sample collection. Alkalinity was determined by strong acid Gran titration at a field laboratory near Raquette Lake, New York, within one week of sample collection. Samples for analyses of major cations (Ca^{2+}, Mg^{2+}, Na^+, K^+, NH_4^+), anions (SO_4^{2-}, NO_3^-, Cl^-) and silicic acid were shipped to the University of Virginia where analyses were performed as described by Galloway *et al.* (1984). Dissolved organic carbon analyses were performed by Cronan (1985).

Samples for Al, Fe, and fluoride determinations were shipped to Cornell University for analysis. All Al measurements were made by spectrophotometric analysis, using modifications of the Ferron-orthophenanthroline method described by Rainwater and Thatcher (1960). Specific modifications were as follows:

(1) A mixed and aged (5 days minimum) color developing reagent was used to obtain lower and more stable blanks (Bersillon *et al.*, 1980).

(2) Sample and reagent volumes were scaled down by a factor of ten for automated reagent additions and analysis in a Bausch & Lomb Spec-400 Automated Spectro-photometer System.

(3) Elapsed time between addition of color developing reagents and absorbance measurement at 370 nm was standardized at 160 seconds.

(4) Corrections for organic color absorbance were obtained by measuring absorbance of samples plus a special mixed reagent prepared by substituting distilled water for the Ferron solution. These color blank absorbance measurements were then subtracted from the Al determination absorbances.

Total acid reactive aluminum (Alr), total monomeric aluminum (Ala), and organic monomeric aluminum (Alo) were determined as described by Driscoll (1984). The determination of the latter fraction (Alo) was slightly modified as follows:

(1) Rexyn (101) cation exchange resin (5 cm^3) was packed in 20 cm polypropylene colums and titrated with 0.007 M NaCl and 0.005 M HCl to give pH 5.5, 0.001 M NaCl eluant.

(2) Sample volumes of 45 mL were passed through the colums at a flow rate of 13.3 mL min^{-1} from disposable plastic syringes in a Sage Syringe Pump. The last 20 mL of column effluent was retained for organic Al analysis.

(3) Columns were eluted with 45 mL of 0.001 M NaCl between samples. Columns were replaced after every 50 samples.

Total Digested (Al_T): Sample aliquots were digested by addition of 10 µL of 20% $(NH_4)_2S_2O_8$ to 2.5 mL of sample. The resulting solutions were held at 95 °C for 30 min. Cooled samples were then analyzed for Al by the Ala procedure.

Total Organic (Al_{TO}): Analyses were conducted as above for aliquots of samples passed through the cation exchange columns. Reactive iron (FeR) and total iron (FeT) were determined as described by Rainwater and Thatcher (1960). Fluoride measurements were obtained with an Orion ionspecific electrode. Procedures described in the Orion Instruments electrode manual were used.

3. Results

3.1. ACID/BASE CHEMISTRY OF THE ILWAS LAKES

The average pH, alkalinity, and concentrations of major ions obtained for outlet samples from Woods, Panther, and Sagamore Lakes during the period 1978–1980 are presented

TABLE I

Average annual concentrations in the water leaving the Woods, Panther and Sagamore lakes calculated by averaging weekly samples from the outlets over a 3-yr period (January 1, 1978 to December 31, 1980) (Units are μeq L^{-1} except for pH)

	Woods	Sagamore	Panther
pH [a]	4.7	5.6	6.2
Alkalinity (titrated)	− 10.0	30.6	147
Strong acid anions			
SO_4^{2-}	126	163	123
NO_3^-	19.4	24.1	23.4
Cl^-	9.1	16.6	12.6
C_A (sum of acid anions)	154	204	159
Base Cations			
NH_4^+	3.16	0.68	1.64
Ca^{2+}	73.1	147	207
Mg^{2+}	18.6	55	52
Na^+	19.3	35	41
K^+	6.2	12	12
Al [b]	16.4	2.6	1.5
C_B (sum of base cations)	137	252	315

[a] Converted to H^+-ion concentration before averaging.
[b] Total monomeric aluminum concentration, in μmol L^{-1} times its charge at the equivalence point, i.e. 3.0.

in Table I. The marked differences in pH levels of the three lakes are readily explained by consideration of the following relationship:

$$[H^+] = C_A + \propto Ca - C_B \qquad (1)$$

where

C_A = the equivalent sum of strong acid-forming anions (SO_4^{2-}, NO_3^-, Cl^-);
C_B = the equivalent sum of base-forming cations (Ca^{2+}, Mg^{2+}, Na^+, K^+, and Al); and
$\propto Ca$ = the equivalent sum of dissociated weak acids (principally HCO_3^-).

Hydrogen-ion activity is explicitly defined by the electroneutrality condition specified in Equation (1), and alkalinity or acid neutralizing capacity can be expressed as:

$$ALK = C_B - C_A = \propto Ca - [H^+]. \qquad (2)$$

This equivalent to the approach used in the ILWAS model (Gherini et al., 1985).

The term ($\propto$ Ca – [H$^+$]) in Equation (2) is equivalent to operationally determined Gran's plot alkalinity and the term ($C_B - C_A$) may be obtained by summation of the major ion equivalents. Plots of ($C_B - C_A$) vs titration alkalinity for samples from Panther and Woods lakes (Figure 1) illustrate the equality of these terms. The mathematical equality is presented in the paper on the ILWAS model. (Gherini *et al.*, 1985). The relatively small deviations of the pH-alkalinity data from that expected for an aqueous CO_2 system to which strong acids and bases have been added, indicates that

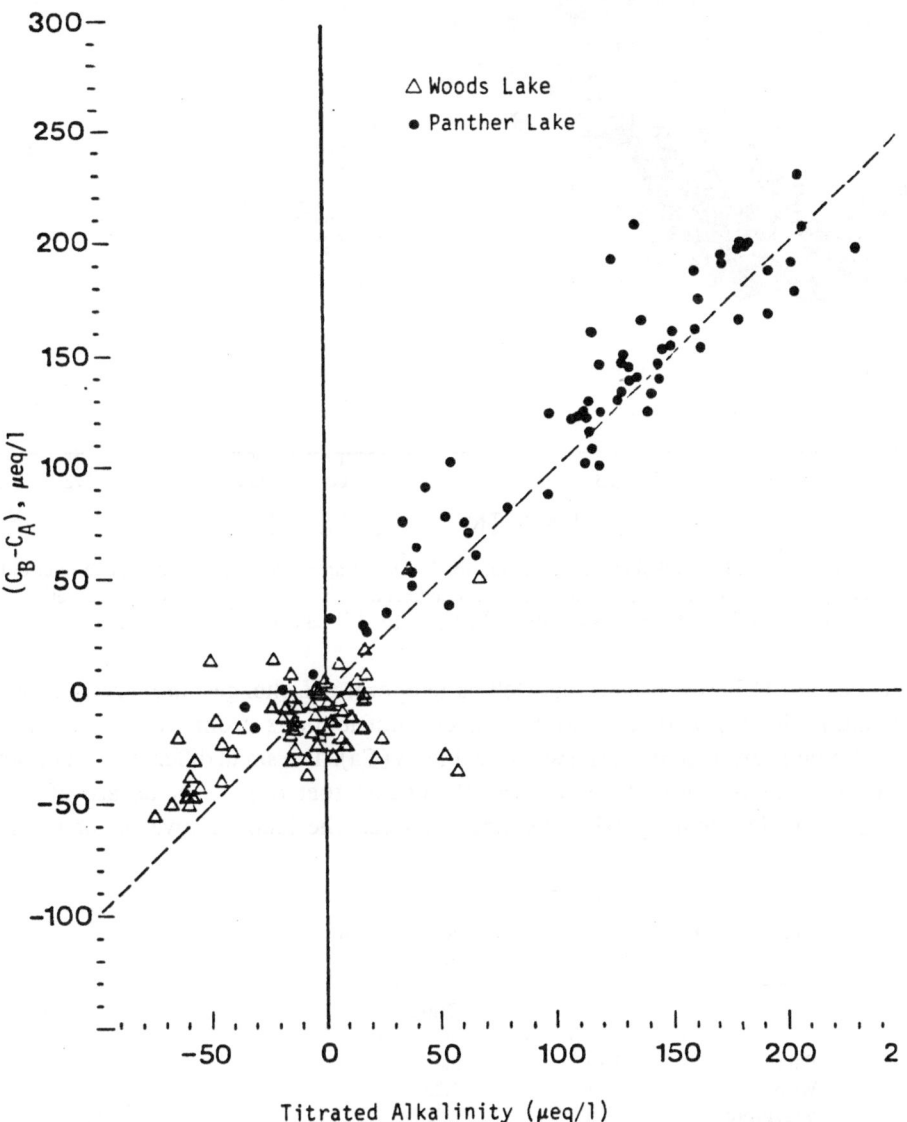

Fig. 1.   Comparison of ($C_B - C_A$) and titrated alkalinity for samples collected from Woods and Panther lakes. Dashed line has a 1 to 1 slope. (C_B = Ca^{2+} + Mg^{2+} + K$^+$ = Na$^+$ + NH$^+$ + 3Al; C_A = SO$_4^{2-}$ + NO$_3^-$ + Cl$^-$).

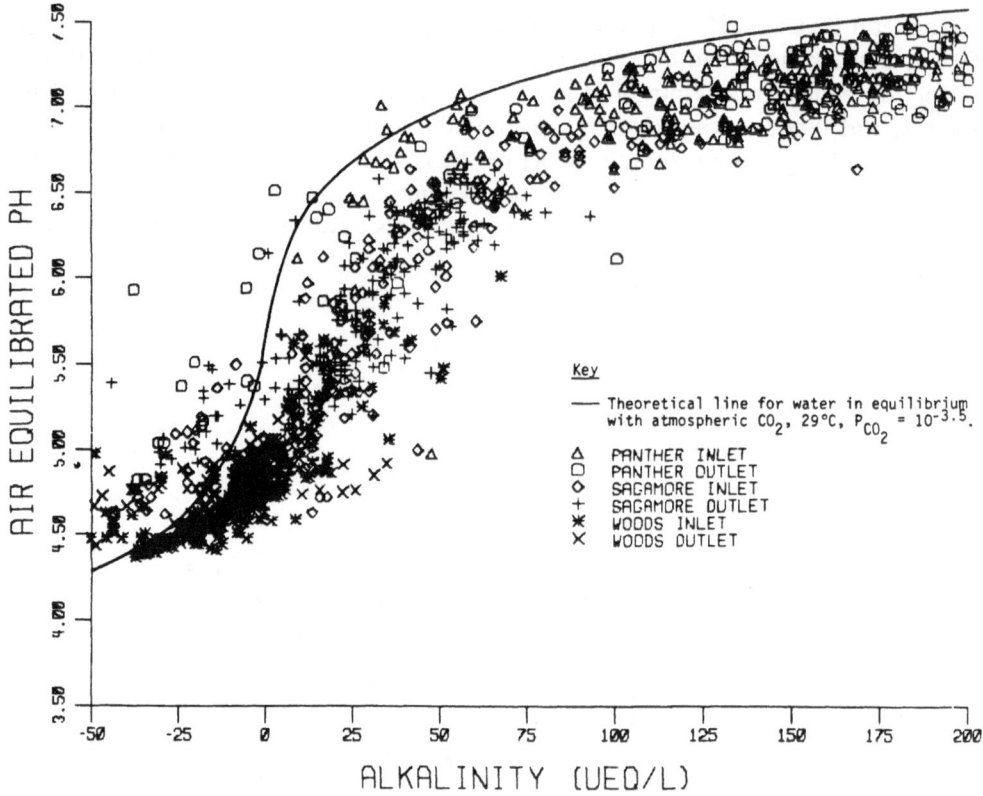

Fig. 2. Air-equilibrated pH-alkalinity relationship for Woods, Sagamore, and Panther Lake inlet-outlet samples compared to theoretical curve for water in equilibrium with atmospheric CO_2. (Theoretical line based on $K_1 = 10^{-6.38}$, $K_2 = 10^{-10.38}$, $K_H = 10^{-1.41}$, and $P_{CO_2} = 10^{-3.5}$ atm.)

differences in pH between lakes, as well as temporal pH changes, are largely a result of variation in C_B and C_A, with smaller contributions from variations in the non-carbonate components of the weak acid term (Ca). The small differences in organic carbon levels of the three lakes (Table II) indicate that organic acids are of lesser importance in determining pH differences between the lakes. However, at the low

TABLE II

Average DOC (mg L^{-1}) levels in the inlets and outlets of the ILWAS basins

	Inlet	Outlet
Panther	3.14	3.47
Woods	4.23	2.25
Sagamore	5.76	4.92

* 15–35 µmol L^{-1} could be present if an average of 7 µmol mg^{-1} DOC is assumed (Cronan, 1985).

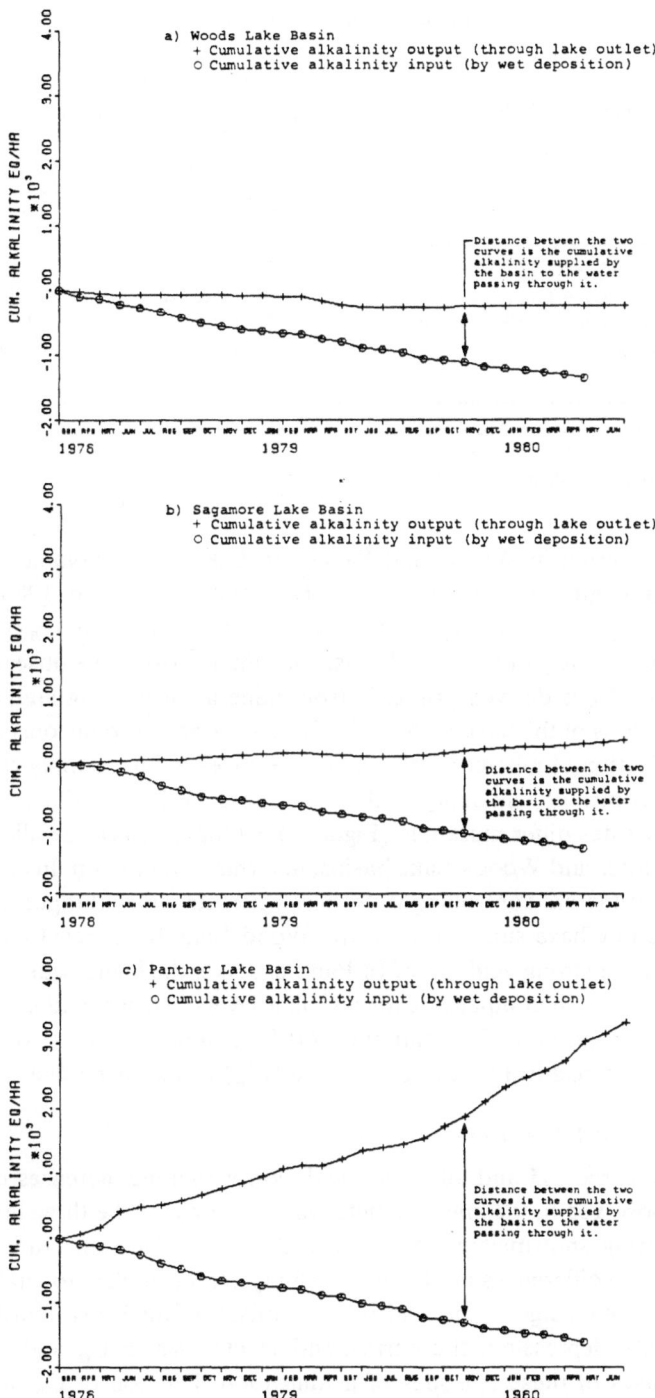

Fig. 3.   Cumulative alkalinity input and output for Woods, Sagamore, and Panther Lake Basins. (a) Woods
Lake Basin, (b) Sagamore Lake Basin, and (c) Panther Lake Basin. Input is by wet deposition; output is
through lake outlets. Vertical distance between the curves in each figure is the cumulative alkalinity supplied
by the basin to the water passing through it. Data sources: RPI, University of Virginia, USGS; calculations
by Tetra Tech.

TABLE III

Major ion fluxes in Panther and Woods basins (Input values are for wet deposition only)

| | eq ha^{-1} yr^{-1} | |
Flux	Panther	Woods
$Alk_{in} = (C_B - C_A)_{in}$	− 1304 ($\underline{H}^+$)	− 1253 (H$^+$)
$Alk_{out} = (C_B - C_A)_{out}$	+ 1530 ($\underline{HCO_3^-}$)	− 133 (H$^+$)
Net Alk supply = $Alk_{out} - Alk_{in}$	2834	1120
Total $C_{B_{out}}$	3430	966
Al_{out}^{n+} [a]	59 (2%)	219 (23%)

Note: $C_B = Ca^{2+} + Mg^{2+} + Na^+ + K^+ + Al^{n+}$ *
$\quad C_A = SO_4^{2-} + NO_3^- + Cl^-$.
[a] Three equivalents per mole of Al.

alkalinity levels present in Woods and Sagamore Lakes, both organic acid* and Al would be predominant buffering components (Driscoll and Bisogni, 1983). Inspection of the major ion data in Table I indicates that the large difference in C_B between lakes, compared to the relatively uniform C_A levels, contribute most to the observed alkalinity differences. Since C_B is derived primarily from mineral weathering and ion exchange processes in the soils of the basins, the differences in concentrations observed between basins are a reflection of inherent watershed differences in base supply rate. Although all three basins contribute alkalinity to the water passing through the systems, the net alkalinity supply rates differ markedly (Figure 3). Comparison of the alkalinity supply rates for the Panther and Woods Lake basins, in terms of major ion fluxes, emphasizes the importance of factors controlling C_B export when strong acid input levels are high (Table III). The low base supply rate in the Wood Lake basin results in incomplete neutralization of the strong acid input. In Panther Lake, the higher rate of base supply is more than sufficient to compensate for a similar level of strong acid input. Note also that a significant proportion of the cation export from Woods Lake is Al (23% of total C_B), compared to a relatively small percentage (2%) in Panther Lake.

3.2. TEMPORAL ACIDIFICATION

Episodes of decreased pH and alkalinity, with corresponding increases in Al concentrations, were observed in the inlets, outlets, and epilimnia of the three lakes under ice cover during spring snowmelt periods (Figures 4 to 6). The same rationale used to evaluate inter-basin differences in pH and alkalinity can be applied to analyze the basis for these short-term changes in he acid/base chemistry of the lakes. During periods of pH and alkalinity depression, the corresponding changes in C_B and C_A (Figure 7) indicate the causes of these episodes of acidification. Dilution of C_B in low-flow by increased snowmelt runoff was associated with increased levels of C_A, primarily from elevated NO_3^- concentrations. The net result is a decrease in alkalinity and corresponding pH drop, following the relationship previously described (Figure 1). Dilution is a natural consequence of snowmelt runoff, but the magnitude of alkalinity depression

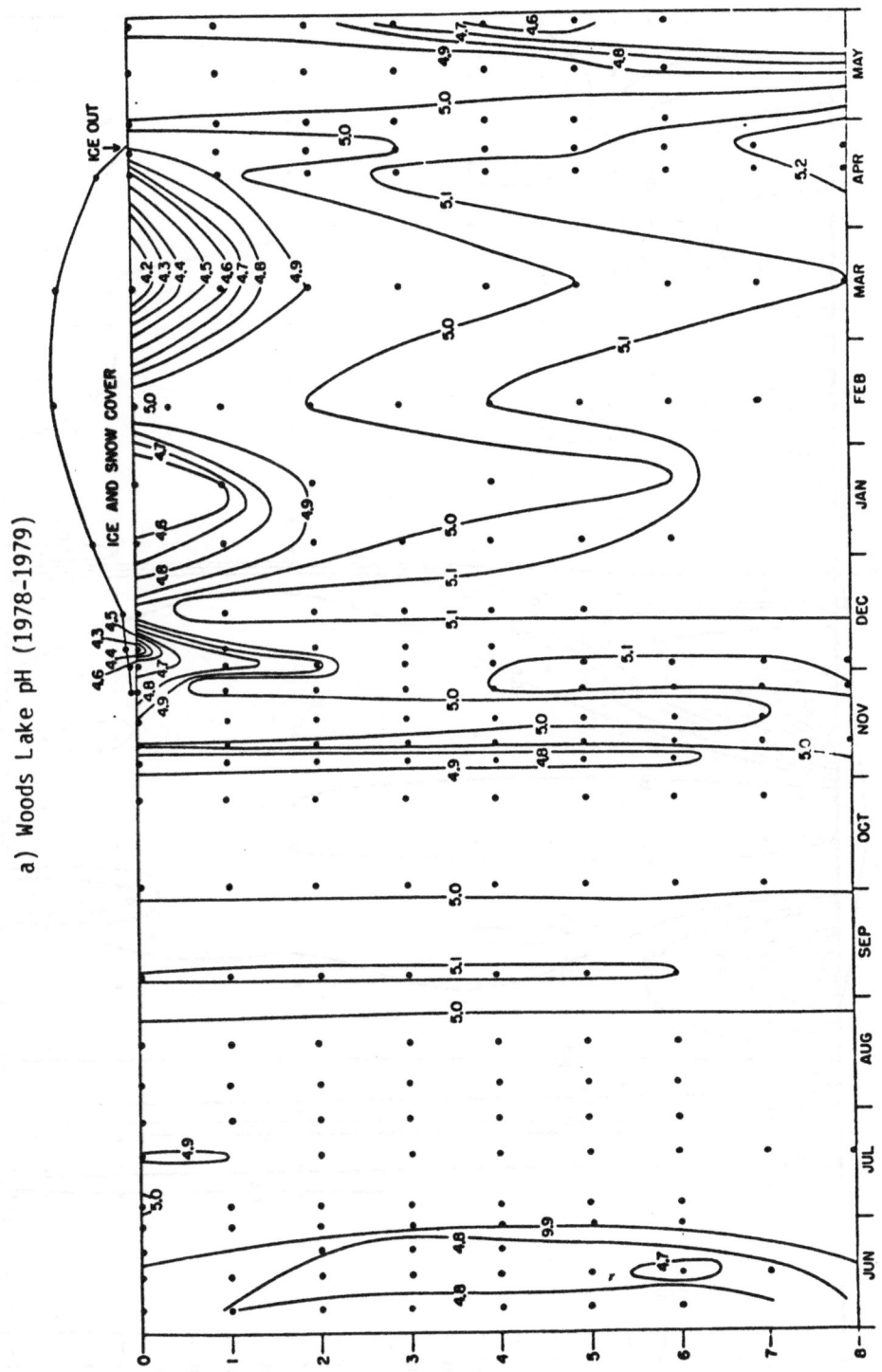

a) Woods Lake pH (1978–1979)

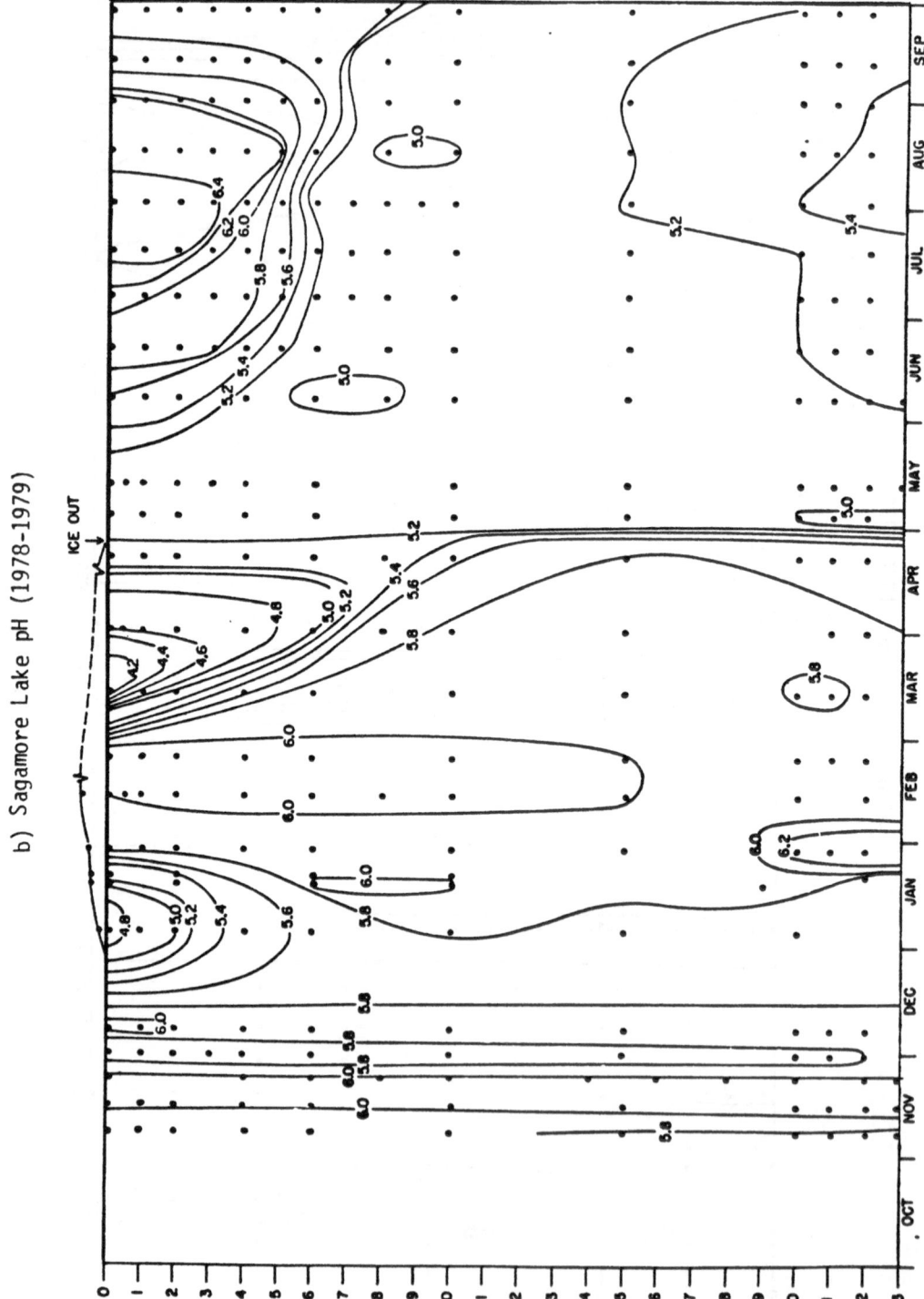

b) Sagamore Lake pH (1978-1979)

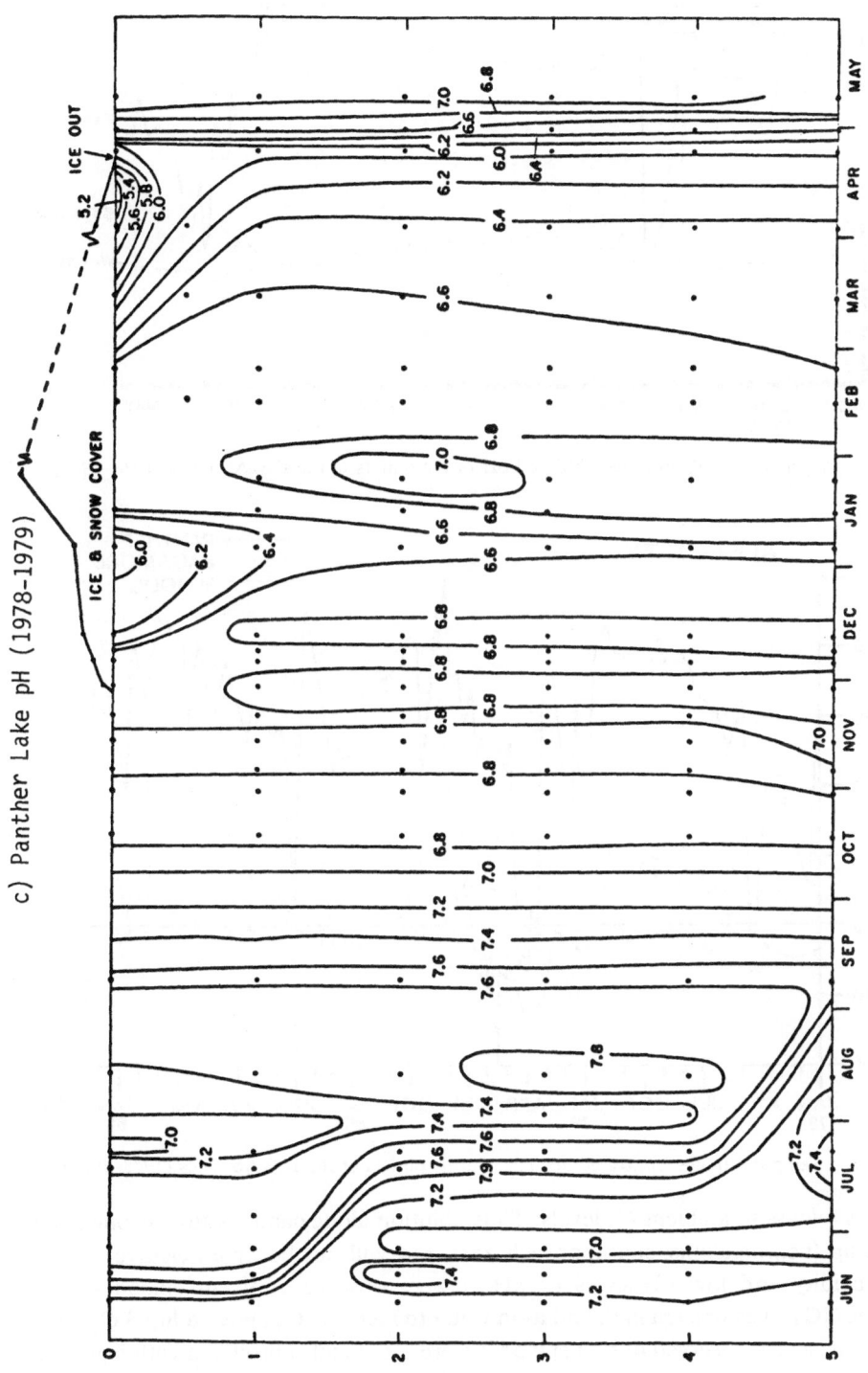

Fig. 4. Isopleths of pH in (a) Woods, (b) Sagamore, and (c) Panther lakes during 1978–1979.

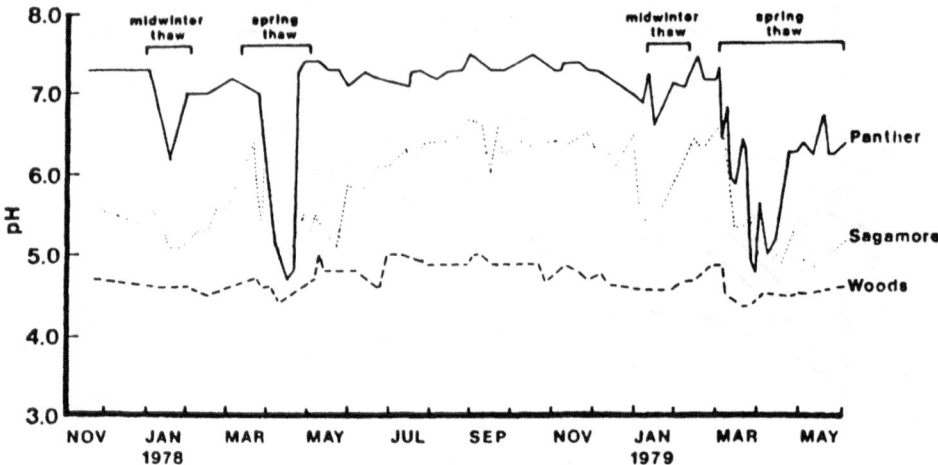

Fig. 5.   Temporal trends in air-equilibrated pH of the outlets of the ILWAS lakes (1978–1979).

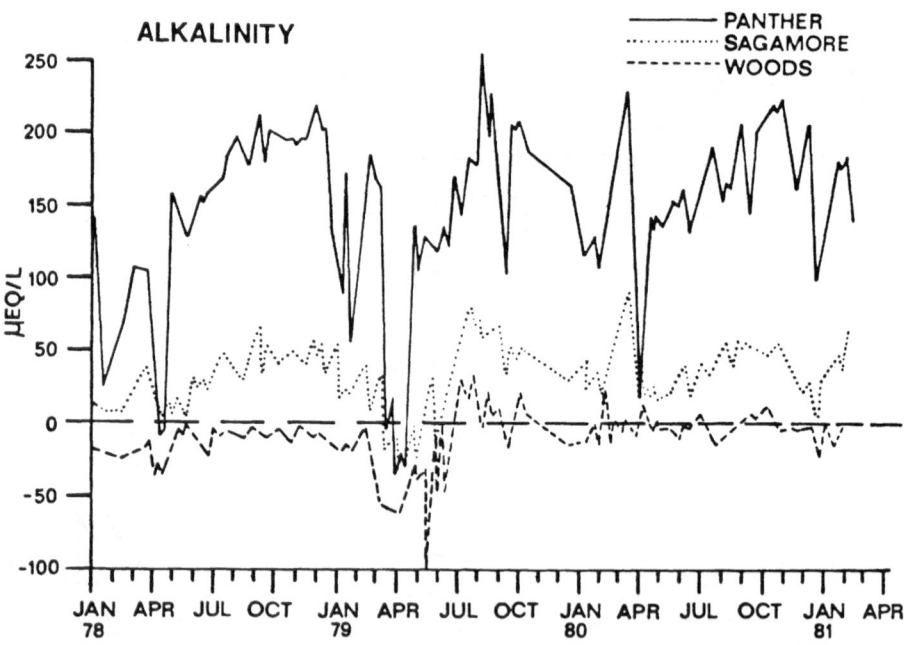

Fig. 6.   Temporal trends in titrated alkalinity in the outlets of the ILWAS lakes (1978–1980).

caused is *relative* to ambient C_A levels. Thus, dilution alone cannot cause strong acidity to develop (i.e., alkalinity can approach zero, but will not become negative). Excess strong acidity and low pH levels ($<$pH 5) were observed only in episodes where sufficient NO_3^- was present in the dilution water to increase C_A above diluted C_B levels. Elevated Al levels observed during episodes were also highly correlated with high NO_3^- concentrations in the Woods and Panther Lake basins (Figure 8).

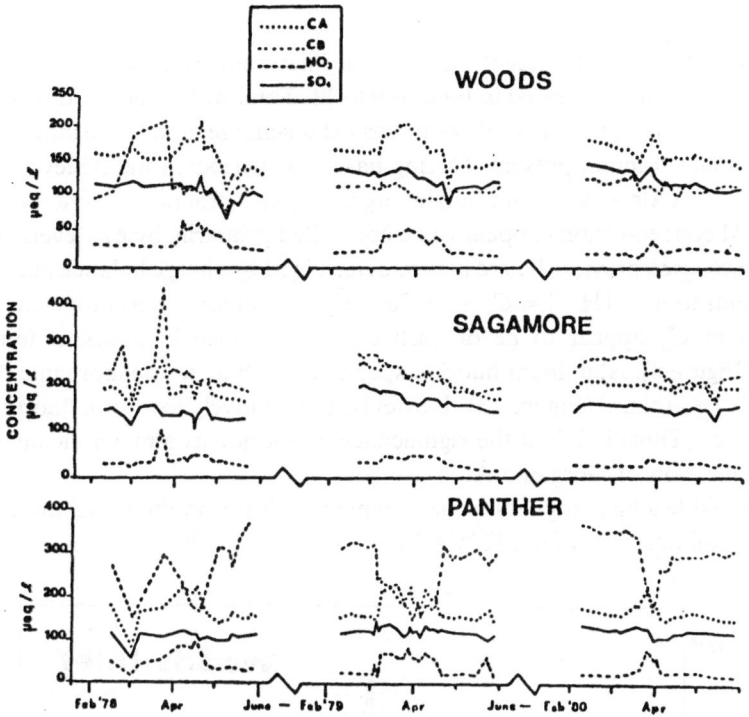

Fig. 7. Temporal trends in alkalinity components in the outlets of the ILWAS lakes (1978, 1979, and 1980).

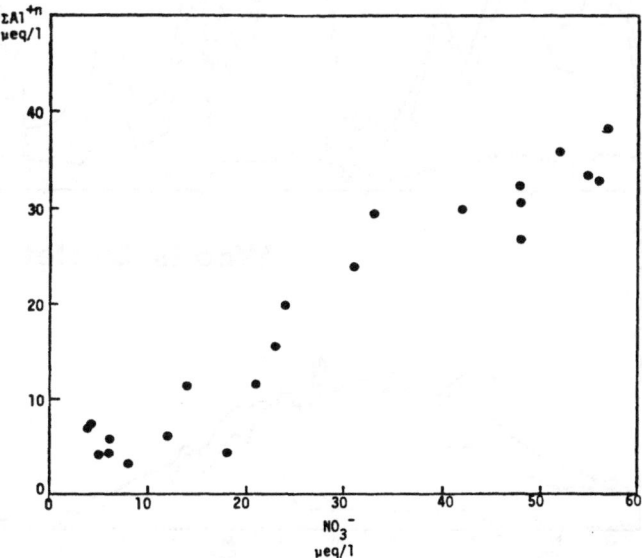

Fig. 8. Comparison of NO_3^- and Al^{n+} equivalents in Woods Lake outlet and inlet samples during snowmelt (1980).

3.3. ALUMINUM CHEMISTRY

Temporal variations in Al concentrations in the inlets and outlets of Woods, Sagamore, and Panther lakes are illustrated in Figures 9 to 11. Total Al concentrations were highest in all the watersheds during periods of increased discharge, and labile inorganic Al was the predominant fraction present. During base flow periods, total Al levels decreased and non-labile (Alr – Ala) and labile organic (Alo) fractions were predominant. Inorganic Al concentrations appear to be controlled primarily by pH levels (Figure 12) which, as discussed previously, are in turn determined by charge balance and total weak acid concentrations ($[H^+] = C_A + \propto Ca - C_B$). Temporal variations in the NO_3^- component of C_A appear to be of particular significance in terms of inorganic Al transport (Figure 8). Aluminum fluoride species contribute to the inorganic Al fraction at pH levels less than 6 (Figure 13). However, the relatively low total fluoride levels in all of the lakes (Table IV) limit the significance of fluoride as a major element involved in the mobilization of inorganic Al.

Organic acid leaching appears to be a significant factor in the initial mobilization of Al in upper soil horizons of the ILWAS basins (Cronan, 1985). However, the mobility

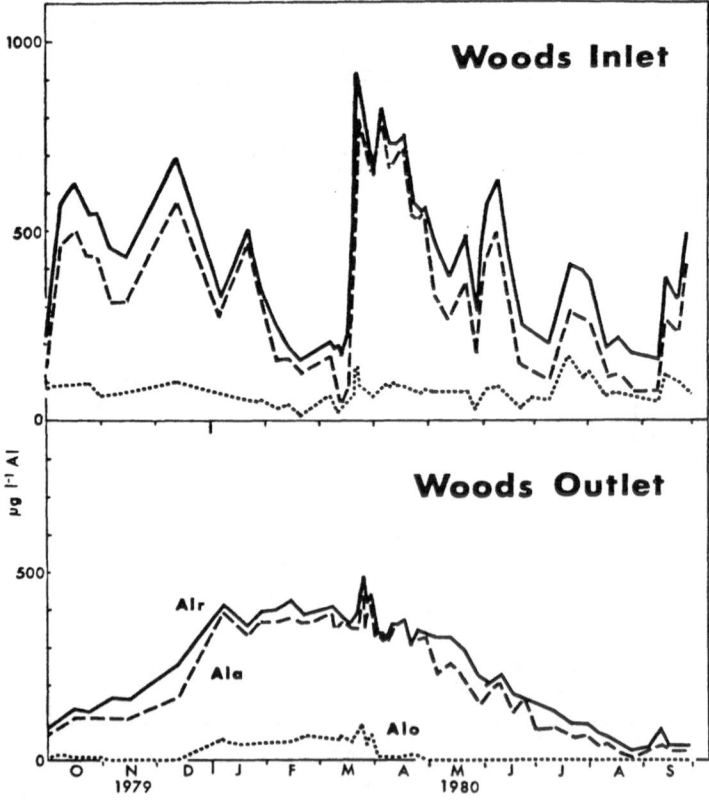

Fig. 9.   Temporal trends in Al fractions in Woods Lake inlet and outlet. Alr = total acid reactive Al; Ala = total monomeric Al; and Alo = organic monomeric Al.

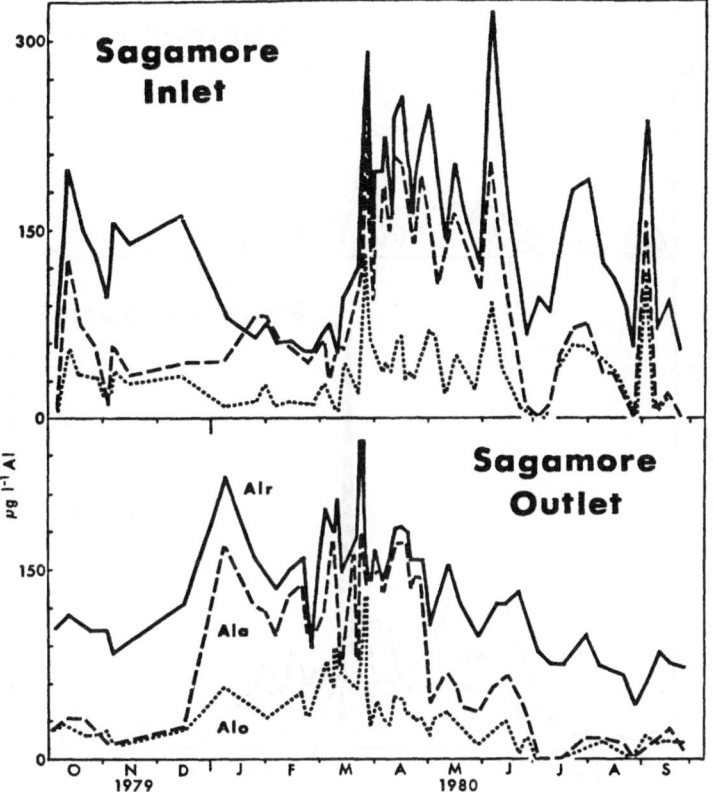

Fig. 10. Temporal trends in Al fractions in Sagamore Lake inlet and outlet. Alr = total acid reactive Al;
Ala = total monomeric Al; and Alo = organic monomeric Al.

TABLE IV

Total fluoride and fluoride species levels in the ILWAS basins (Means and standard
deviations for monthly samples collected during 1981) (Units are $\mu mol\ L^{-1}$)

Station	F^-	HF	ΣAlF	ΣF
Panther outlet	6.55	0.003	0.62	7.17
	(2.44)	(0.004)	(1.89)	(1.41)
Panther inlet	3.59	0.001	0.00	3.60
	(0.78)	(0.003)	–	(0.78)
Sagamore outlet	4.28	0.007	0.74	5.03
	(1.25)	(0.005)	(1.08)	(0.04)
Sagamore inlet	3.83	0.007	1.22	5.06
	(1.65)	(0.008)	(1.63)	(0.72)
Woods outlet	0.28	0.007	2.46	2.75
	(0.31)	(0.006)	(0.29)	(0.28)
Woods inlet	0.14	0.005	3.02	3.17
	(0.17)	(0.003)	(0.74)	(0.74)

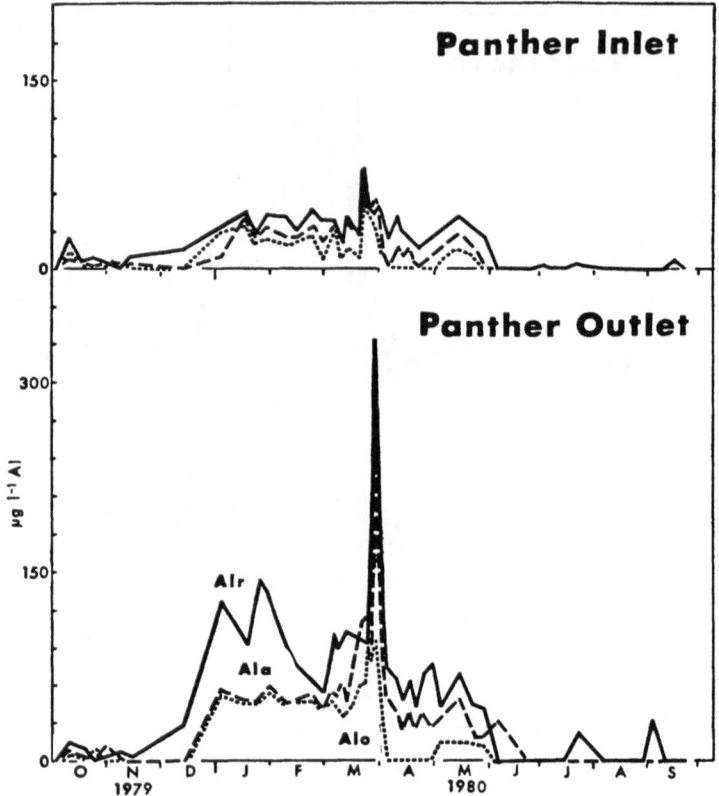

Fig. 11.   Temporal trends in Al fractions in Panther Lake inlet and outlet. Alr = total acid reactive Al; Ala = total monomeric Al; and Alo = organic monomeric Al.

of organically complexed Al appears to be low compared to inorganic Al associated with strong acidity. This relation is also evident from comparison of the relative concentration of the two fractions (Alo and Ala − Alo) in the inlets and outlets of the three basins (Figures 9 to 11). Additionally, a progressive decrease in both organic carbon and Alo was observed when comparing soil solutions, lake inlets, and lake outlets of the three basins (Table V).

TABLE V

Changes in Alo and DOC of soil lysimeter solutions, and inlets and outlets of the ILWAS lakes (Mean concentrations for 1981 samples in mg L^{-1})

Site	Panther		Woods		Sagamore	
	Alo	DOC	Alo	DOC	Alo	DOC
Shallow lysimeter	0.285	23.6	0.203	19.9	0.216	15.9
Deep lysimeter	0.137	6.1	0.126	5.4	0.064	5.8
Lake inlet	0.003	3.1	0.090	4.2	0.039	5.8
Lake outlet	0.007	3.5	0.012	2.2	0.030	4.9

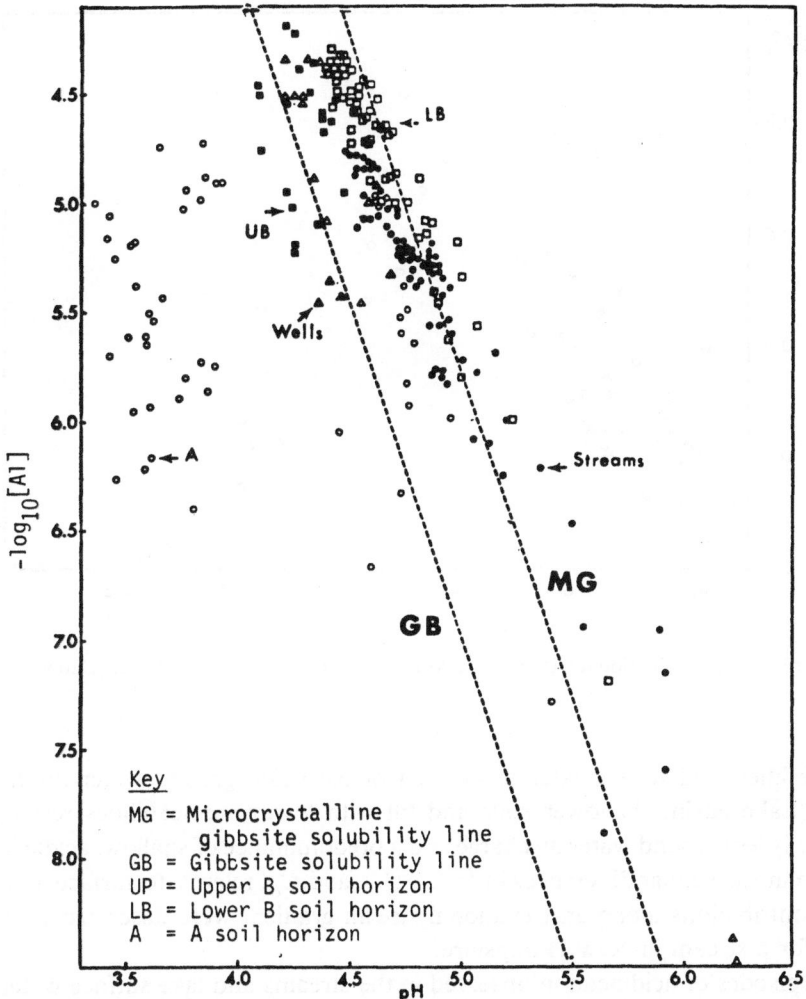

Fig. 12. Labile inorganic Al ($-\log_{10}$ concentration) for ILWAS soil water and stream samples in relation to pH.

4. Discussion

The differences in alkalinity (and hence, pH) observed between Woods, Sagamore, and Panther lakes were found to result primarily from the different base cation supply rates in each watershed, relative to similar strong acid input levels. The resultant differences in net alkalinity supply to the surface waters of these basins appear to be controlled primarily by soil characteristics and hydrologic features of the watersheds. As discussed (in the other papers in this issue), flow paths and residence time of water in the soils are primary determinants of the relative degrees of aqueous acidification or cation release resulting from strong acid inputs. Thus, in Panther Lake, deep tills and high soil

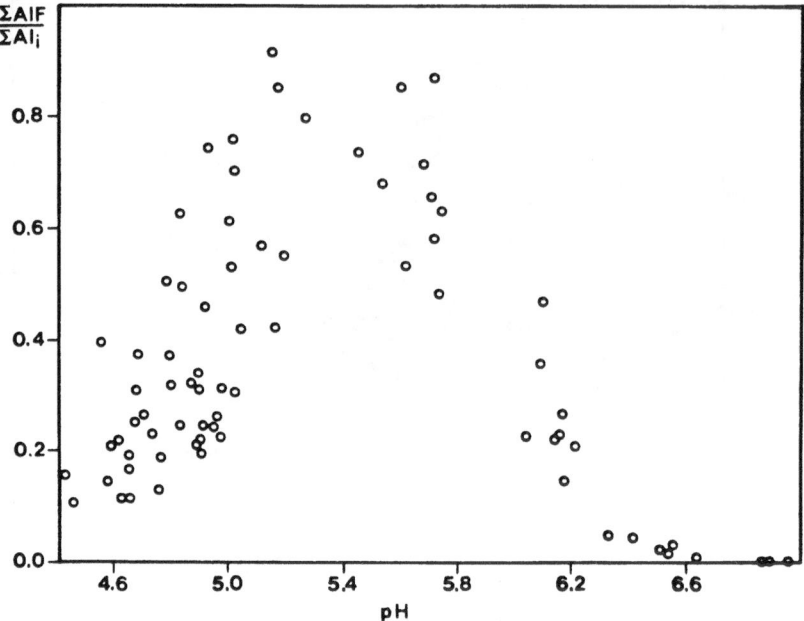

Fig. 13.   Total Al fluoride/total inorganic Al ratios in ILWAS lake samples in relation to pH.

permeabilities lead to a greater proportion of base rich ground water discharge. In Woods Lake basin, shallower soils and till with lower permeabilities result in proportionally less ground water discharge and a predominance of shallow, acidic interflow enriched in Al. Similar differences in H^+, Al^{n+}, and C_B, relative to surface and ground water contributions along an elevational stream gradient were described by Johnson (1979) for a stream in New Hampshire.

The episodes of acidification observed in the streams and lake surface water during snowmelt are also related to the same structural and hydrologic basin characteristics described above. The observed dilution of C_B, which was particularly pronounced in Panther Lake, and the corresponding increase in C_A are related to an upward shift in flowpaths from ground water dominated base flow to shallow interflow during increased snowmelt discharge. The elevated inorganic Al levels and corresponding increases in nitrate observed during these episodes suggest that HNO_3 derived from the snowpack and nitrification in the soils triggers acidification and Al mobilization. The magnitude of this effect is tempered by the degree of base flow C_B dilution relative to ambient sulfate levels. These observations support the hypothesis that strong acid neutralization is a two-stage process, with initial Al mobilization in upper soil horizons, followed by primary mineral dissolution and alkalinity production in deeper soil horizons. The separation of these processes in time and/or space results in incomplete neutralization, acidification, and export of inorganic Al to surface waters. Similar conclusions were reached by Driscoll and Schafran (1984); and Johnson (1979).

Aluminum export rates were less than 2% of total base cation output from Panther Lake basin and 23% of base cation output from Woods Lake basin. The similarly of soil chemistry in the upper horizon of both Woods and Panther basins and the comparable strong acid input levels suggest that initial Al mobilization rates in the soils of both basins should be similar and considerably greater than the observed total aluminum exports in outflow (Panther: $0.79 \, \text{kg ha}^{-1} \, \text{yr}^{-1}$; Woods: $2.25 \, \text{kg ha}^{-1} \, \text{yr}^{-1}$). The greater depth and higher permeability of soils in the Panther basin suggest that re-precipitation in soil weathering zones is a primary sink for Al mobilized in upper soil horizons by both strong and organic acids. An undetermined proportion of the base cation export could thus be attributed to the neutralization of Al acidity, although it is probably minor compared to carbonic acid driven base cation release rates. Lower soil permeabilities and shallower flow paths in the Woods Lake basin appear to account for the greater rates of Al export. However, the marked differential observed in total Al levels between tributary inflow and lake outflow together with a decrease in organic carbon and Al, suggest that Woods Lake itself serves as a secondary sink for part of the Al exported from the soils of the watershed during the summer months. Monthly fluxes of Al into the lake from the watershed were calculated from tributary inflow concentrations and surface runoff estimates (Peters, 1985). When compared to the measured fluxes of Al leaving the lake via outflow (Figure 14), marked seasonal) trend in apparent Al sedimentation is observed. During the winter period of ice cover (4 to 5 mo), only 27% of the Al entering the lake is retained. However, retention of Al in the lake increases to 63% of the input during the summer. The increased pH in the lake during the summer probably triggers the precipitation of Al. This displacement of Al accumulation zones from the terrestrial soil horizons (spodosol B horizons) to lake sediments may be a relatively recent phenomenon associated with increased Al mobility resulting from excess strong acid inputs. Higher concentrations of acid

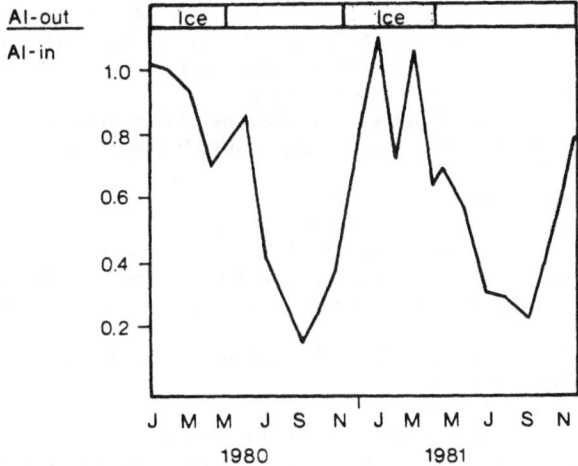

Fig. 14.   Monthly ratios of Al outlet flux to estimated input flux from the watershed of Woods Lake.

extractable Al in surface sediments, compared to deeper sediments in Wood Lake, support this hypothesis (Galloway and Schofield, 1982).

Acidification of surface waters can be defined as a decrease in alkalinity $(C_B - C_A)$. On a watershed basis, the alkalinity of the water passing through the ecosystem is determined by the relative rates of acid input (from both atmospheric deposition and internal generation) and base supply (primarily from soil weathering and ion-exchange processes). Results from this study have shown that interbasin differences in base supply rates and temporal dilution of base inputs, relative to strong acid levels, are the primary determinants of net alkalinity supply. Increased atmospheric inputs of strong acid could be expected to cause acidification if the rates of input exceed watershed base supply rates. Simulations with the ILWAS model indicate that base supply rates increase with increased atmospheric loadings to partially offset added acidity. Land use changes or watershed distrubances that might decrease base supply rates relative to atmospheric strong acid input levels could also lead to acidification. Extensive fires or logging are examples of watershed disturbances that could alter base supply rates by increasing rates of cation uptake and storage in the forest during regeneration. There are no records of fires or extensive logging activity in the Panther Lake watershed from 1890 to the present. The only intensive harvest was for shoreline spruce salvaged after severe storm blow-down damage in 1950. There are also no records of fires in the Woods watershed, although several large burns were recorded in the surrounding area in 1903 and 1908. Extensive harvest of softwoods took place from 1914–1924, followed by selected cutting of hardwoods in 1950, 1970, and 1975. The net effect of these disturbances in forest growth on the base supply rate from the Woods Lake watershed is unknown, but probably are less significant than atmospheric acid inputs in terms of acidbase balance (Gherini et al., 1985).

References

Baker, J. P. and Schofield, C. L.: 1982, *Water, Air, and Soil Pollut.* **18**, 289.

Bersillon, J. L., Hsu, P. H., and Fiessinger, F.: 1980, *J. Soil Sci. Soc. Amer.* **44**, 630.

Cronan, C.: 1985, *Water, Air, and Soil Pollut.* **26**, 335 (this issue).

Driscoll, C.: 1984, *Internal. J. Environ. Anal. Chem.* **16**, 267.

Driscoll, C. and Bisogni, J. J.: 1983, 'Weak Acid/Base Systems in Dilute Acidified Lake and Streams in the Adirondack Region of New York State', in J. Schnoor (ed.), *Modeling of Total Acid Precipitation Impacts*, Ann rbor Sci. pp. 55–73.

Driscoll, C. and Schafran, G.: 1984, *Nature* **310**, 308.

Galloway, J. N. and Schofield, C. L.: 1982, unpublished data.

Galloway, J. N., Altwicker, E. R., Church, M. R., Cosby, B. J., Davis, A. O., Hendrey, G., Johannes, A. H., Nordstrom, K. D., Peters, N. E., Schofield, C. L., and Tokos, J.: 1984, *The Integrated Lake-Watershed Acidification Study*, Volume 3: Lake Chemistry Program. Electric Power Research Institute, Palo Alto, CA.

Gherini, S. A., Chen, C. W., Mok, L., Goldstein, R. A., Hudson, R. J. M., and Davis, G. F.: 1985, *Water, Air, and Soil Pollut.* **26**, 425 (this issue).

Johnson, N. M.: 1979, *Science* **204**, 497.

Peters, N. E.: 1985, *Water, Air, and Soil Pollut.* **26**, 387 (this issue).

Pfeiffer, M. and Festa, P.: 1980, *Acidity Status of Lakes in the Adirondack Region of New York in Relation to Fish Resources*, FW-P168 (10/80), New York State Dept. of Env. Conserv., Albany, NY.

Rainwater, F. and Thatcher, L.: 1960, *Methods for Collection and Analysis of Water Samples.* U.S.G.S. Water Supply Paper 1454, 297 pp.

Shofield, C. L.: 1976, *Acidification of Adirondack Lakes by Atmospheric Precipitation: Extent and Magnitude of the Problem*, Final Rep. D. J. Proj. F-28-4, NYS Dept. Env. Cons., 11 pp.

Schofield, C. L. and Trojnar, J. R.: 1980, 'Aluminum Toxicity to Brook Trout (*Salvelinus fontinalis*) in Acidified Waters', in Toribara, T., Miller, M., and Morrow, P. (eds.), *Polluted Rain*, pp. 341–365, Plenum Press, NY.

THE ILWAS MODEL: FORMULATION AND APPLICATION

STEVEN A. GHERINI, LINGFUNG MOK, ROBERT J. M. HUDSON,
GEORGE F. DAVIS

Tetra Tech., Inc., 3746 Mt. Diablo Boulevard, Suite 300, Lafayette, CA 94549, U.S.A.

CARL W. CHEN

Systech., Inc., 3744 Mt. Diablo Boulevard, Suite 101, Lafayette, CA 94549, U.S.A.

and

ROBERT A. GOLDSTEIN

Electric Power Research Institute, 3412 Hillview Avenue P.O. Box 10412, Palo Alto, CA 94303, U.S.A.

(Received November 1, 1984; revised May 14, 1985)

Abstract. The Integrated Lake-Watershed Acidification Study (ILWAS) model was developed to predict changes in surface water acidity given changes in the acidity of precipitation and dry deposition. The model routes precipitation through the forest canopy, soil horizons, streams and lakes using mass balance concepts and equations which relate flow to hydraulic gradients. The physical-chemical processes which change the acid-base characteristics of the water are simulated by rate (kinetic) and equilibrium expressions and include mass transfers between gas, liquid and solid phases. The aqueous constituents simulated include: pH, alkalinity, the major cations (Ca^{2+}, Mg^{2+}, K^+, Na^+, and NH_4^+) and anions (SO_4^{2-}, NO_3^-, Cl^-, F^-), monomeric Al and its inorganic and organic complexes, organic acid analogues and dissolved inorganic carbon (C_T). Since free hydrogen ion (H^+) (and hence pH) is not conserved, its concentration is derived from the solution alkalinity and the total concentrations of inorganic C, organic acid, and monomeric Al.

The ILWAS model has been used to predict changes in the acidity of Woods Lake (typical lake pH 4.5 to 5.0) and Panther Lake (typical lake pH 6 to 7) given reductions in total atmospheric S loads. The two basins are located within 30 km of each other in the Adirondack Mountains and receive similar acidic deposition. The response to a halving in the total atmospheric S load was basin-specific: In Panther Lake, little pH change occurred even 12 yr after the load reduction; in Woods Lake, the change was considerably larger.

Hypothesis testing with the model has shown that the routing of water through soils (shallow versus deep flow) largely determines the extent to which incident precipitation is neutralized. Analysis of the two lake basins using the model and field data showed the watersheds to be net suppliers of base to the through-flowing water, although internal watershed production of strong acidity did occur. This internal production of acidity was approximately two-thirds the amount of the atmospheric load.

1. Introduction

The ILWAS model was developed to predict changes in surface water acidity given changes in deposition acidity incident to forested ecosystems. Specifically, the model was developed to predict surface water H^+ and Al concentrations because of their importance to fish. As conceptualization work progressed, it became obvious that the model must simulate both the routing of water to the lakes and the major alkalinity producing and consuming reactions occurring along the flow pathways. The result was a unified theory of lake acidification quantified by equations and algorithms embedded in the model code.

Water, Air, and Soil Pollution **26** (1985) 425–459. 0049–6979/85.15.

PART A. MODEL THEORY, FORMULATIONS, AND SOLUTION TECHNIQUES

The spatial heterogeneity in lake-watershed systems is represented in the ILWAS model by a network of homogeneous compartments. In essence, the ILWAS model routes water from one compartment to another and calculates the concentrations of dissolved constituents by simulating the biogeochemical reactions taking place in each compartment. To accommodate lake-watershed heterogeneity, the hydrologic basin is divided into subcatchments, stream segments (if streams are present) and a lake (Figure 1a). In each subcatchment, there are compartments to represent the canopy, snowpack, and soil layers (Figure 1b). The lake is divided into horizontally mixed layers to allow for calculation of temperature and water quality profiles.

To simulate lake water pH, the model follows a set of 16 aqueous chemical constituents using mass balance equations (See Table I). The change in the concentration of these species in any compartment results from equilibrium and rate-limited source and sink processes and dispersive and advective inputs and outputs across the compartment boundaries. The compartments together are treated mathematically as a series of mixed reactors (CSTR's) as is shown in Figure 2. H^+-ion is not one of the 16 mass-balanced constituents. Its concentration is derived from the solution alkalinity (acid-neutralizing capacity) and analytical total concentrations for inorganic carbon (C_T), analogue organic acid (R_T), and monomeric aluminum (Al_T).

The original conceptualization of the ILWAS model, its formulation, and application are described elsewhere (Goldstein et $al.$, 1984; Chen et $al.$, 1979, 1983). Details of the

TABLE I

Solution phase chemical constituents simulated by the ILWAS model[a]

Cations	Anions	Uncharged Species	Analytical Total	Gases (deposition only)
Ca^{2+}	SO_4^{2-}	H_4SIO_4	Alk [ANC]	(SO_x)
Mg^{2+}	NO_3^-	$CO_2(Aq)$	Org acid 1	(NO_x)
K^+	Cl^-	AlF_3	Org acid 2	
Na^+	$H_2PO_4^-$	AlR_1	Al_T	
NH_4^+	F^-	$Al(R_2)_3$	C_T (DIC)	
H^+	$Al(OH)_4^-$	$Al(OH)_3^0$	F_T	
Al^{3+}	HCO_3^-	H_3R_1		
$Al(OH)^{2+}$	CO_3^{2-}	HR_2		
$Al(OH)_2^+$	R^{3-}			
AlF^{2+}	HR_1^{2-}			
AlF_2^+	$H_2R_1^-$			
$Al(SO_4)^+$	R_2^-			
AlR_2^{2+}	AlF_4^-			
$Al(R_2)_2^+$	AlF_5^{2-}			
	AlF_6^{3-}			
	$Al(SO_4)_2^-$			

[a] Constituents shown in italics are derived from the others.

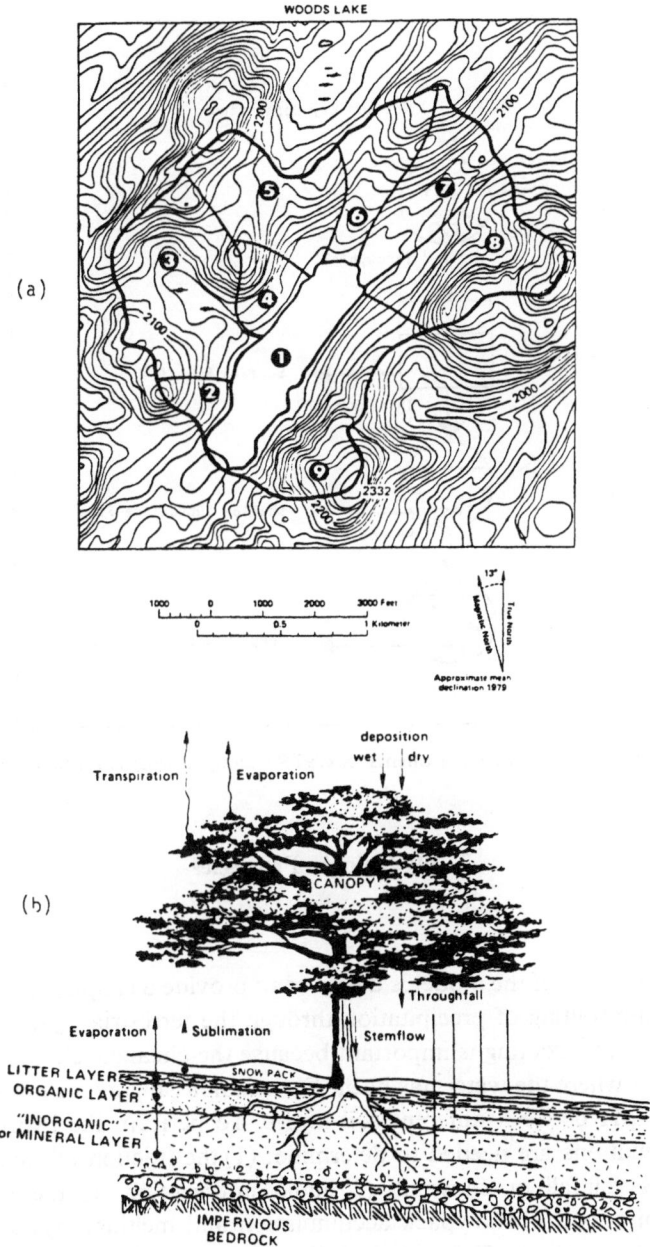

Fig. 1. Dividing a watershed into homogeneous compartments: (a) hydrologic basin subcatchments; (b) subcatchment compartments.

model are presented in the ILWAS report series, Volume 1: *Model Principles and Application Procedures* (EPRI EA-3221, V1). The following section reviews model theory, formulations, and solution techniques.

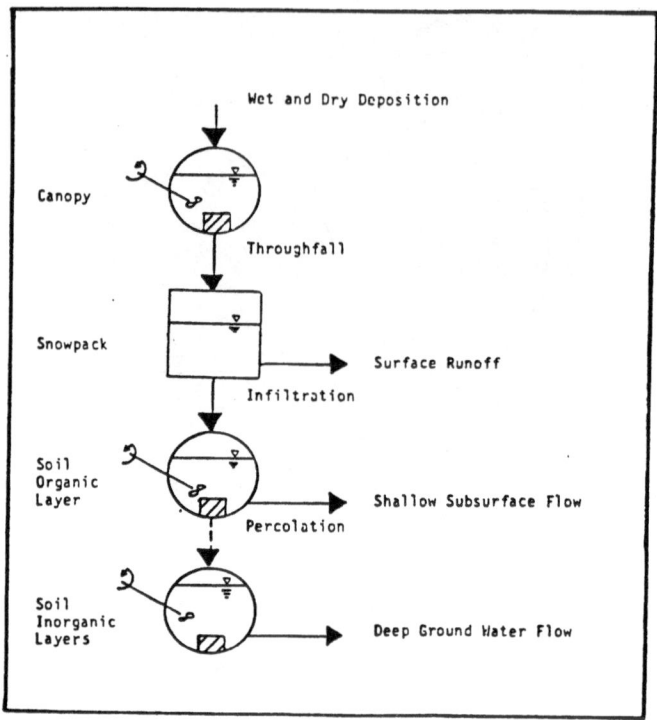

Fig. 2.   Idealization of system compartments as CSTR's along flowpaths through a subcatchment.

2. Hydrologic Processes

2.1. THEORY

The hydrologic portion of the model is designed to provide a simple, yet representative, description of the routing of precipitation through the terrestrial system and into the surface waters. Water routing is important because the chemical characteristics of the water depend on where the water has been. For example, Figure 3 shows the variation in soil solution pH between different soil layers. To perform the hydrologic routing calculations, equations are needed to determine: (1) the fraction of incoming precipitation which is rainfall or snowfall, (2) liquid interception by the forest canopy, (3) evapotranspiration, (4) snowpack accumulation and melting, (5) flow through the soil layers, (6) stream flow, (7) the vertical distribution of lake inflow and withdrawal of outflow, and 8) lake volumetric discharge.

Three types of hydrologic equations are used in the model. The mass conservation principle is used to make water balances for every compartment of the system. Rate equations are used to relate water flow to a potential driving force. For example, Darcy's Law is used to relate percolation and lateral flow in a soil layer to hydraulic gradient, and stage-flow relationships are used to relate lake discharge to lake surface elevation.

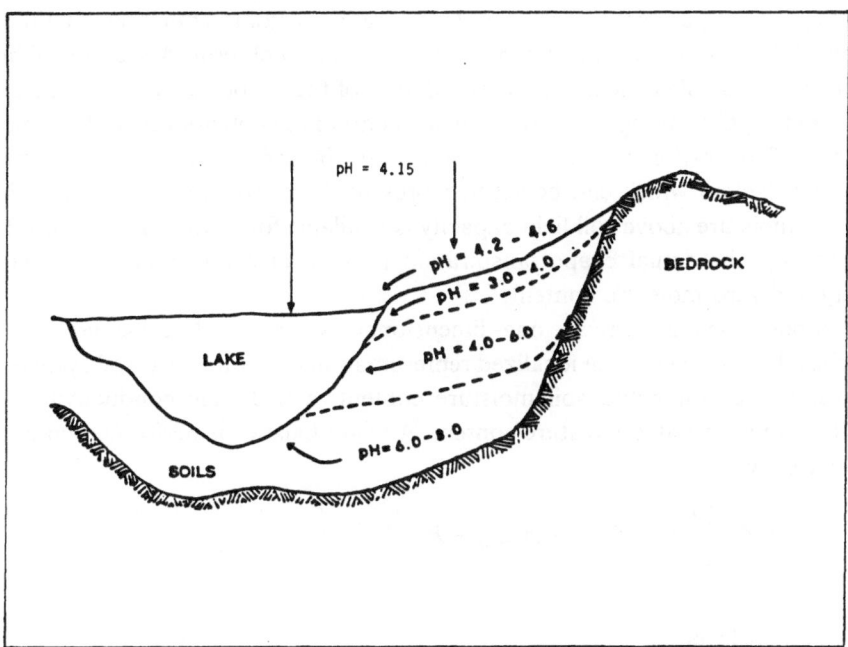

Fig. 3.   Variation in soil solution pH. The depth to which precipitation seeps into the soil before flowing laterally influences the lake water pH.

The model also uses empirical correlations to estimate hydrologic variables such as evapotranspiration. The following section will highlight the major formulations used.

2.2. Formulations

Precipitation can be in the form of rain, snow, or exist as a mixture of rain and snow. The model determines the snow fraction of precipitation based upon ambient air temperature. Snowfall is assumed not to be intercepted by the canopy. A canopy water interception capacity is calculated from monthly leaf area indices and a maximum interception storage depth. If rainfall is greater than interception capacity, the excess water becomes throughfall.

Accumulation of snowpack is calculated by mass balance. The model allows snowpack to dissipate by sublimation and temperature-induced and rain-induced melting. The sublimation rate is assumed constant but can differ for open areas and forested areas. Melting rate coefficients and a correction factor for the temperature-induced snow melting computation are input calibration parameters. After snowmelt rates are calculated, the snowmelt water must first satisfy the field capacity of the snowpack before draining. The snowpack 'field capacity' is the amount of liquid water the pack can retain by capillarity.

The daily potential evapotranspiration is determined by the product of an evapotranspiration factor, the mean air temperature, a humidity correction factor and a calibration scaling parameter. The evapotranspiration factor is a function of latitude and is composed of two terms, a constant term and a seasonal term. A second calibration parameter is available to scale the seasonal term of the evapotranspiration factor. The canopy interception storage is used to satisfy a part of the potential evapotranspiration. The unsatisfied portion is taken from soil layers in accordance with user-specified distribution factors which can be set to represent the distribution of roots in the soil layers. All moisture above soil field capacity is available for evapotranspiration. Below field capacity, the actual evapotranspiration is reduced exponentially between field capacity and zero moisture content.

The model uses a vertical, one-dimensional system to describe ground water hydraulics. Figure 4 shows an idealized representation of a watershed soil system. Each soil layer has a volumetric soil moisture content, a hydraulic conductivity, a field capacity, and a saturated moisture content. A water balance is performed on each soil layer as follows:

$$A_j Z_j \frac{d\theta_j}{dt} = A_j P_{j-1} - A_j E_{pj} - P_j A_{j+1} - L_j \qquad (1)$$

where

A_j = area of layer j;
Z_j = the thickness of layer j;
θ_j = the average volumetric water content of layer j;
A_{j+1} = the surface area of layer $j + 1$;
P_{j-1} = the percolation flux from the layer above ($j - 1$) into layer j;
E_{pj} = actual evapotranspiration flux from layer j;
P_j = the percolation flux from layer j to the layer below ($j + 1$) and;
L_j = the lateral flow rate flux from layer j.

The model assumes that the percolation rate is zero at and below field capacity and increases linearly with soil moisture content to the hydraulic conductivity at saturation. The lateral flow is the product of the hydraulic conductivity, width of the soil layer, hydraulic gradient, and an effective saturated depth. The effective saturated depth is zero at field capacity and increases linearly with soil moisture content to the depth of the soil layer at saturation. These formulations of vertical and lateral flow are mathematically equivalent to a Darcian approach with the unsaturated pressure gradient term neglected and the unsaturated hydraulic permeability assumed to vary linearly with soil moisture between saturation and field capacity.

In the model, Equation (1) is set up for each soil layer. The boundary conditions are such that P_{j-1} is the infiltration flux for the very top layer and P_j is zero for the bottom layer, using an impervious bedrock boundary condition. Since Equation (1) is not coupled between layers, the mass balance equation can be solved sequentially from the top down. Iterative corrections are made in overlying layers if P_j exceeds the potential for the next layer to receive water.

Water can accumulate on the soil surface when the entire soil column is saturated, or when the saturated hydraulic conductivity limits the infiltration rate. The surface

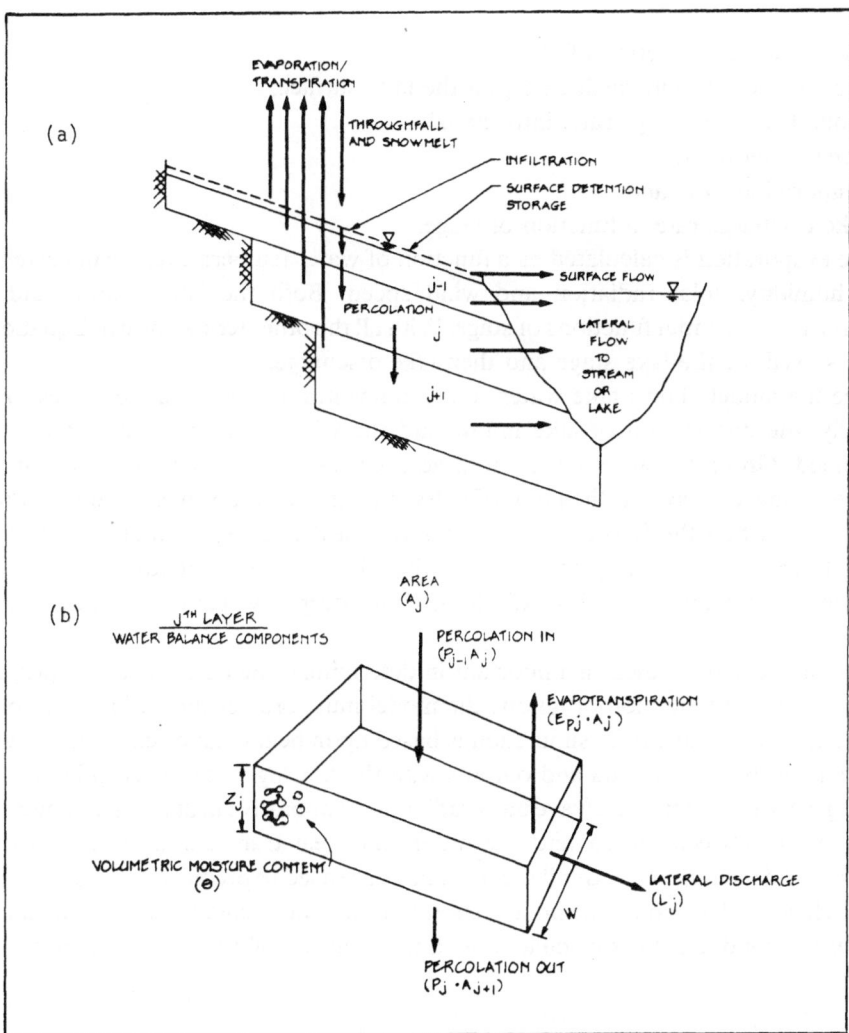

Fig. 4.   Hydrologic representation of a watershed soil system: (a) flow routing; (b) water balance components of a single soil layer.

runoff and surface detention storage are determined by solving a water balance equation and Manning's free surface flow equation simultaneously.

Stream hydrology follows Muskingum routing. After the outflow and the change in storage are computed for each stream segment, the stream depth is calculated from stage cross-sectional area relationships.

The lake discharge is computed from an overall water balance on the lake:

$$\frac{dV}{dt} = P + G + S - E - O \qquad (2)$$

where

V = lake volume, a function of stage,

P = precipitation or snowmelt rates, on the lake surface,

G = ground water seepage rate, into the lake,

S = stream inflow rate,

E = evaporation rate, and

O = lake discharge rate, a function of stage.

Lake evaporation is calculated as a function of water temperature, mean air temperature, humidity, solar radiation and wind speed. Both the lake volume and lake discharge are user-input functions of stage. With all the other terms known, Equation (2) can be solved for the lake stage and then lake discharge.

Since the model allows lake water solute concentrations and temperatures to vary vertically, the distribution of lake inflow and the withdrawal of outflow have to be determined. Ground water seepage from adjacent land catchments is assumed to be mixed with the lake layer at the point of entry. Stream flow, however, is routed into the lake at a level where the density of the stream water is the same as that of the lake. Water flowing to the outlet comes primarily from the lake surface. Depending on the degree of density stratification and lake discharge rate, deeper water is entrained into the outflow.

Since the density gradient is important in determining the outflow withdrawal zone and the vertical distribution of inflow, the model must predict the temperature profile for the lake. The temperature simulation is based upon heat balance calculations which include advective heat inputs and outputs with the flowing water, absorption of short and long wave radiation, long wave back radiation, evaporative heat transfer, conductive heat transfer between the air and the water, and heat transfers associated with the formation and melting of ice. Similar calculations are used to predict the soil temperature profile. Heat is advected by the infiltrating water and conducted by the soil media. The soil thermal conductivity is calculated as a function of soil moisture content.

3. Chemical Processes

3.1. THEORY

In the preceding section on hydrologic processes, the underlying theory was based upon two principles: conservation of mass (COM) and flow down energy gradients. The COM principle is used again for tracking the chemical solutes along with a conservation of charge principle. A fundamental problem, though, occurs with the use of the COM principle alone here. The major chemical constituent of interest, the free H^+ concentration, or pH, cannot be directly predicted by simple mass balance calculations. Whereas, when two Na^+ solutions are mixed in equal volumes, the resulting Na concentration is just the average of the two, this rarely occurs for H^+. Why is this? There are two reasons:

(1) The simple analytical procedure for H^+ concentration (the pH meter and the glass electrode) only measures the concentration of H^+ which is free in solution – the H^+ associated or complexed with other species (e.g., weak acids) is not measured. This is *not* the case for the other major ions of interest. Typically, measurements of Ca, for example, include not only the free ion, Ca^{2+}, but also $Ca(OH)^+$, $Ca(OH)_2^0$, $CaCl^+$, etc.

(2) The H^+ associated with the other species, such as the weak acids, dissociate to a degree dependent upon the existing H^+ concentration in solution. Equilibria of the form $HB \rightleftarrows H^+ + B^-$ exist.

3.2. ALKALINITY

The computational difficulty associated with the nonconservative behavior of the free H^+-ion is overcome by using an extended form of the solution alkalinity relationship. Alkalinity is a measurable parameter which is used frequently in water treatment calculations to determine how much acid or base must be added to change the free H^+-ion concentration (or pH) of a solution. Alkalinity is a conservative parameter and can be used directly in mass balance equations. Alkalinity is routinely measured by titration with strong acid to an equivalence point (inflection point); several analytical procedures have been developed to help identify this point (e.g. Gran, 1952).

The alkalinity of a water is its acid-neutralizing capacity. The higher the alkalinity of a water, the more strong acid it takes to reduce the pH to a fixed value. This ability to resist pH decreases derives from aqueous species which neutralize H^+-ions as they are added and thereby offset the increase in free H^+-ion concentration. For example, bicarbonate ion, HCO_3^-, the major acid-neutralizing species in most surface waters, combines with hydrogen ions to form aqueous carbon dioxide:

$$HCO_3^- + H^+ \rightarrow CO_2(aq) + H_2O . \tag{3}$$

The alkalinity of a water can be defined as the concentration of all such H^+-ion acceptors minus the free H^+-ion concentration (Stumm and Morgan, 1970).

$$Alk = [HCO_3^-] + 2[CO_3^{2-}] + [OH^-]$$
$$+ [\text{other } H^+\text{-ion acceptors}] - [H^+] \tag{4}$$

where
Alk   = alkalinity (acid-neutralizing capacity);
$[HCO_3^-]$ = bicarbonate concentration;
$[CO_3^{2-}]$ = carbonate concentration;
$[OH^-]$ = hydroxide concentration;
$[H^+]$ = free H^+-ion concentration; and
$[\ ]$ = molar concentrations .

For most waters, all but the first term on the right-hand side of Equation (2) is negligible and can be disregarded. In low alkalinity waters, however, the concentration of the 'other H^+-ion acceptors' often becomes large relative to the total concentration of bicarbonate, carbonate, and hydroxide. These other H^+-ion acceptors include organic substances with carboxyl and phenolic hydroxyl groups, for example:

$$R\text{-}COO^- + H^+ \rightarrow R\text{-}COOH \text{ (organic acids)} \tag{5}$$

and monomeric Al species and their complexes, for example,

$$Al(OH)_2^+ + 2H^+ \rightarrow Al^{3+} + 2H_2O \tag{6}$$

$$Al \cdot R + 3H^+ \rightarrow Al^{3+} + H_3R. \tag{7}$$

The extended alkalinity in the ILWAS model includes water itself, the carbonate system, the monomeric Al system and its organic complexes, and dissolved organic carbon (DOC). The DOC alkalinity is represented alternatively by a triprotic and/or monoprotic model organic acid with fixed dissociation constants and a fixed number of acid-base functional groups per unit mass of carbon (μeq/mg C^{-1}). The components of the total alkalinity, as represented by the H^+-ion acceptors, are given below:

$$Alk = Alk_{H_2O} + Alk_C + Alk_{R_1} + Alk_{R_2} + Alk_{Al} + Alk_{Al \cdot 0} \tag{8}$$

where

$$Alk_{H_2O} = [OH^-] - [H^+] \tag{9}$$

$$Alk_C = [HCO_3^-] + 2[CO_3^{2-}] \tag{10}$$

$$Alk_{R_1} = [H_2R_1^-] + 2[HR_1^{2-}] + 3[R_1^{3-}] \tag{11}$$

$$Alk_{R_2} = [R_2^-] \tag{12}$$

$$Alk_{Al} = [Al(OH)^{2+}] + 2[Al(OH)_2^+] + 3[Al(OH)_3^0] +$$
$$+ 4[Al(OH)_4^-] \tag{13}$$

$$Alk_{Al \cdot 0} = 3[Al \cdot R_1] + [AlR_2^{2+}] + 2[Al(R_2)_2^+] +$$
$$+ 3[Al(R_2)_3]. \tag{14}$$

The proton reference level (Stumm and Morgan, 1970) here includes $CO_2(aq)$, Al^{3+} (e.g. as from $AlCl_3$), H_3R_1, and HR_2. Adding these species to the solution does not change the alkalinity although the pH may change. Different reference levels can also be defined and used by the model.

An alternative representation of solution-phase alkalinity, which is mathematically equivalent to that above, and which is also used in the model, is derived below. The derivation is based upon the alkalinity definition and the solution electroneutrality condition. The derivation is for a metal which hydrolyzes, $M(OH)_{(z-n)}^{n-}$, and a strong acid, HA. This approach can be readily extended to yield a general form for the alkalinity equation. The solution charge balance gives

$$[H^+] + z[M^{z+}] + (z-1)[M(OH)^{(z-1)}] + \ldots + [M(OH)_{z-1}^+] =$$
$$= [OH^-] + n[M(OH)_{(z+n)}^{n-}] + \ldots + [A^-]. \tag{15}$$

From the general alkalinity definition (alkalinity = concentration of the H^+-ion acceptors minus the concentration of H^+-ion donors) (Stumm and Morgan, 1970), we have

$$Alk = [OH^-] + [M(OH)^{(z-1)+}] + \ldots + (z-1)[M(OH)^+_{(z-1)}] +$$
$$+ z[M(OH)_z] + \ldots + (z+n)[M(OH)^{n-}_{(z+n)}]$$
$$+ \ldots - [HA] - [H^+]. \tag{16}$$

Solving the charge balance (Equation (15)) for $[OH^-] - [H^+]$ and substitution into the alkalinity definition (Equation (16)) yields:

$$Alk = -[HA] - [A] + z[M^{z+}] +$$
$$+ (z-1)[M(OH)^{(z-1)+}] + [M(OH)^{(z-1)+}] + \ldots +$$
$$+ [M(OH)^+_{(z-1)}] + (z-1)[M(OH)^+_{(z-1)}] + z[M(OH)_z] + \ldots +$$
$$+ (z+n)[M(OH)^{n-}_{(z+n)}] - n[M(OH)^{n-}_{(z+n)}]. \tag{17}$$

Letting

$$M_T = [M^{z+}] + [M(OH)^{(z-1)+}] + \ldots + [M(OH)^+_{(n-1)}] +$$
$$+ [M(OH)_z] + \ldots + [M(OH)^{n-}_{(z+n)}] \tag{18}$$

and

$$A_T = [HA] + [A^-] \tag{19}$$

yields

$$Alk = z \cdot M_T - A_T. \tag{20}$$

Expansion to multiple base and acid systems yields

$$Alk = \sum_i^n |z_i| M_{T_i} - \sum_j^m |z_j| A_{T_j}. \tag{21}$$

This equation relates alkalinity to the difference between the sum of the total non-hydrogen cations (free and complexed forms) times the individual uncomplexed cation charges at the equivalence point (ΣC_B, meq L^{-1}) minus the sum of the total strong acid anions (free and complexed) times the individual uncomplexed anion charges at the equivalence point (ΣC_A, meq L^{-1}) (i.e., $Alk = \Sigma C_B - \Sigma C_A$). Equation (21) can be written in condensed form,

$$Alk = \sum_{k=1}^1 z_k N_k \tag{22}$$

where N_k is the analytical (total) concentration of constituent k. Note that z_k is determined by the charge of the uncomplexed constituent at the equivalence point (alkalinity titration inflection point).

Alkalinity is a conservative parameter, independent of temperature, pressure, and the partial pressure of CO_2 for a solution not reacting with the solid phase. Although these factors change the relative sizes of the terms in the alkalinity equation, they do not change the sum. For example, increasing the partial pressure of CO_2 increases the H^+ concentration but also increases the concentration of bicarbonate and carbonate in a directly offsetting manner and thereby yields no net change in alkalinity,

$$CO_2 + H_2O \rightleftharpoons H^+ + HCO_3^- \rightleftharpoons 2H^+ + CO_3^{2-} . \tag{23}$$

Transport and mass balance equations can be readily written for alkalinity, which is then used in the determination of the H^+ ion concentration.

3.3. CALCULATION OF H^+ CONCENTRATION (OR pH)

The H^+ concentration is determined implicitly from the alkalinity and the analytical sums of total inorganic carbon, C_T, total aqueous aluminum, Al_T, and total organic acid, R_T (all of which are traced by the model using direct mass balance procedures), i.e.,

$$H^+ = f(\text{Alk}, C_T, , R_T) \tag{24}$$

where
$C_T = [CO_2(aq)] + [HCO_3^-] + [CO_3^{2-}]$;
$Al_T =$ the total monomeric Al concentration in solution, including the complexes with
$\quad\quad OH^-$, F^-, SO_4^{2-}, and organic acid ligand (see Table I); and
$R_T =$ the total organic acid in solution, including complexes,
$\quad\quad = R_{1_T} + R_{2_T}$.

An expression for alkalinity is prepared in which the individual alkalinity components are all represented by their analytical totals (e.g., C_T) and ionization and complexation constants. For example, the concentration of uncomplexed HCO_3^- is represented by

$$[HCO_3^-] = C_T\left(\frac{[H^+]}{K_1} + 1 + \frac{K_2}{[H^+]}\right)^{-1} \tag{25}$$

where K_1 and K_2 are the first and second acid dissociation constants for the carbonate system. When all such representations are substituted into the alkalinity expression (Equation (8) plus (9) to (14)), an equation results in which $[H^+]$ is the only unknown. The resulting equation is solved by numerical techniques discussed later.

4. Process Representations

The ILWAS model must include the major chemical processes which add acid or base to the water flowing through the basins. The code accounts for acid-base production and consumption by stoichiometric equations which quantify the amount of alkalinity produced or consumed by each reaction. For example, the stoichiometry for the weathering of potassium feldspar, an important acid-neutralizing mineral, indicates that for every mole of feldspar which reacts, 1 mol of H^+ is withdrawn from solution and is replaced by 1 mol of K^+:

$$KAlSi_3O_8 + H^+ + 4\tfrac{1}{2}H_2O \rightarrow \tfrac{1}{4}Al_4(Si_4O_{10})(OH)_8 + K^+ + 2H_4SiO_4 .$$
$\text{(K feldspar)} \quad\quad\quad\quad\quad\quad \text{(kaolinite)} \tag{26}$

Both fast and slow reactions are simulated by the ILWAS model. Fast reactions are here defined as those reactions which go to 90% completion within hours. These reactions are represented by equilibrium expressions and reaction quotients which give the ratio of the reaction product concentrations to reactant concentrations. For example, the hydrolysis of Al^{3+},

$$Al^{3+} + H_2O \rightleftarrows Al(OH)^{2+} + H^+ \qquad (27)$$

is represented by its reaction quotient as follows:

$$\frac{[H^+][Al(OH)^{2+}]}{[Al^{3+}]} = K_{Al_1}. \qquad (28)$$

This reaction comes to equilibrium quickly. Times does not enter into the equation. The interest here is only in the distribution of products and reactants.

Slow reactions are those which do not rapidly approach equilibrium. These reactions are represented by mathematical expressions (rate expressions) which describe how quickly the reaction is proceeding. For example, the weathering of K feldspar, as shown in Equation (26), is slow and is represented by the rate expression:

$$-\frac{dM}{dt} = kM[H^+]^a \qquad (29)$$

where

$-dM/dt$ = rate of decrease in feldspar mass;

k = the specific reaction rate constant;

M = mass of K feldspar;

$[H^+]$ = the H^+ concentration; and

a = power dependency of the reaction rate on the concentration of H^+ (typically 0.3 to 0.7).

All rate-limited processes are represented in a similar fashion although the forms of the rate expression may vary with the type of reaction. Included are saturation kinetics and mass-action-limited approaches to equilibrium.

Solution of the equilibrium and rate equations is handled in a unique fashion, based upon the model's representation of the hydrologic basin as a coupled series of mixed reactors (CSTR's). More will be presented on this later. How the chemical processes are represented in the individual basin components is discussed in the following sections.

4.1. FOREST CANOPY PROCESSES

Processes occurring on the canopy modify the incident precipitation. The canopy enhances the collection of dry deposition (SO_2, NO_x, and particulates) and accumulates foliar exudation. The dry deposition collection rate for each chemical species is proportional to the leaf area index, ambient air quality, the species deposition velocity, and a collection efficiency. The foliar exudation rate is proportional to the chemical composition of the leaves, a chemical species amplification factor, and the leaf area index.

SO$_2$ and NO$_x$ deposited on the canopy are assumed to be rapidly converted to sulfate and nitrate. Ammonia on the canopy is oxidized to nitrate at a temperature dependent rate proportional to the mass of ammonia present. The overall imbalance between the production of base cations and strong acid anions produces a net change in alkalinity.

The moisture on the canopy plus that of the rain is mixed. The solution pH is calculated as described previously from the mixture alkalinity and C_T, with the solution being in equilibrium with atmospheric CO$_2$. Throughfall is produced to the extent the volume of this solution exceeds the canopy interception storage.

4.2. SNOWPACK PROCESSES

Sequential snowmelt data have shown that snowmelt waters have much higher solute concentrations during the initial phase of the melt. The concentrations decrease exponentially as the melt proceeds (Johannes et al., 1985). This behavior is represented by a simple leaching equation which states that each volume of snowmelt carries away its own solute content and leaches a fraction of the solute from the remaining snowpack.

4.3. SOIL PROCESSES

In terms of neutralizing acidic deposition, the quantitatively most important processes that occur in forested watersheds occur in the soils. The representation of these processes is the most complex portion of the ILWAS model and requires the majority of the computational effort.

The soil is divided into layers which are represented as a series of coupled, mixed reactors (CSTR's) with solid, liquid, and gas phases in each. A schematic summarizing the process within a soil layer is presented in Figure 5. To simulate these processes, the model must account for the flow of water and gases through the soil layer and the processes which change the concentration of constituents associated with the three phases. Many of the important acid-base processes transfer chemical species between the liquid and the soil solids.

4.4. LITTERFALL AND DECAY

Litter from the canopy is deposited onto the upper soil layer at rates which vary monthly and with the type of canopy. A small fraction of the litter-associated chemical species is immediately released to the solution. The remaining litter undergoes four stages of decay:

$$\text{Litter} \xrightarrow{k_1} \text{Fine Litter} \xrightarrow{k_2} \text{Humus and Base Cations}$$

$$\text{Humus} \xrightarrow{k_3} \text{Organic Acid} + \text{aNH}_4^+ + \text{bSO}_4^{2-} + \text{cH}^+ + \text{dCO}_2$$

$$\text{Organic Acid} \xrightarrow{k_4} \text{eNH}_4^+ + \text{fSO}_4^{2-} + \text{gH}^+ + \text{hCO}_2 . \tag{30}$$

Each stage of decay is represented by rate expressions with first order dependencies on the reactant mass or concentration. The specific reaction rate constants, k_i, are

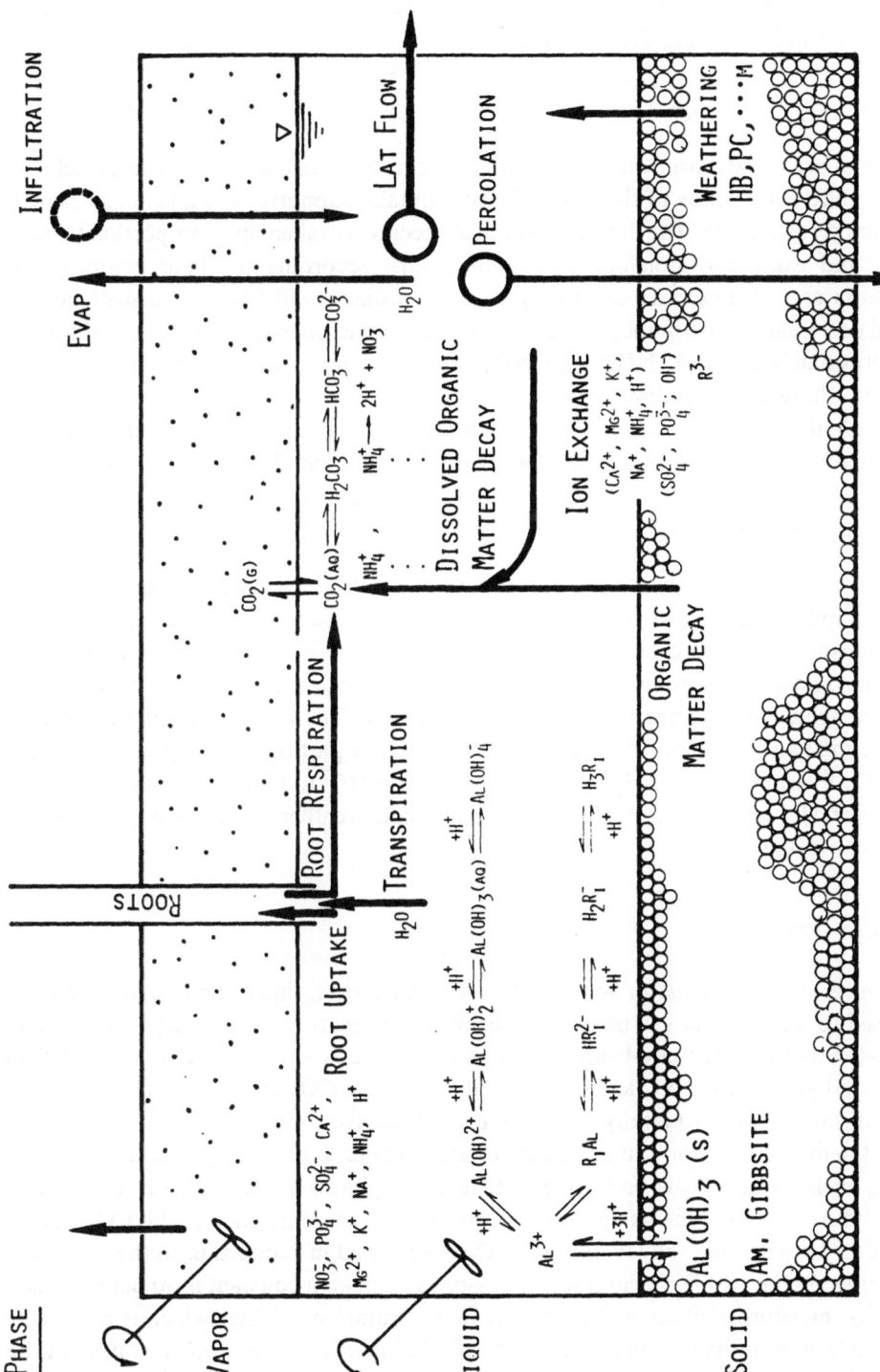

Fig. 5. Idealized soil layer showing major processes.

temperature-dependent and the stoichiometric coefficients, 'a' through 'h', are specific to the type of litter being decomposed.

4.5. NUTRIENT UPTAKE

Nutrients are withdrawn from the soil solution at rates which are set by the model user for the different months of the year. Nutrient uptake supports bole and foliar growth and canopy exudation. The different chemical species are taken up in proportion to bole, foliar, and exudation stoichiometries which vary according to stand composition. Nutrients are withdrawn from soil layers in accordance with the root distribution or model user defined distributions. Nutrient insufficiencies cause all nutrient uptake to be proportionately reduced. NH_4^+ and NO_3^- are taken up in direct proportion to their solution phase concentrations.

The alkalinity of the soil solution is changed by the difference between the uptake of base cations and strong acid anions as given by Equation (21).

4.6. ROOT RESPIRATION

Root metabolic processes release CO_2 in addition to that released by microbial decay. In the model, the release of CO_2 by roots includes fractions associated with basal metabolism and active growth. The latter release occurs in proportion to the monthly nutrient uptake. The production of CO_2 within a subcatchment with a mixed canopy is the weighted sum of production for each tree type. The CO_2 produced dissolves in proportion to its soil gas partial pressure (Henry's Law partitioning). Upon dissolution, the CO_2 hydrates ($CO_2 + H_2O \rightleftarrows H_2CO_3 \rightleftarrows H^+ + HCO_3^-$) and provides an internal source of H^+-ions which can drive exchange and weathering reactions and release alkalinity to the solution.

4.7. CO_2 EXCHANGE

Carbon dioxide can exist in soils in all three phases-gas, liquid, and solid. It can be released to the liquid or gas phases by decay processes, root metabolic processes, and by dissolution of carbonate-bearing rock. It can be transported between layers in both liquid and gas forms. To calculate the total amount of CO_2 in the solution in any soil layer, it is necessary to quantify its inputs and outputs from that layer. This is accounted for in the model as follows: mass balances are made on both the liquid and gas phases within each layer; CO_2 is partitioned between the phases according to Henry's Law; CO_2 diffuses through the gas phase in proportion to the difference in the CO_2 partial pressures between the soil layers; and CO_2 is advected in proportion to the change in moisture content within the layers. The gas-phase diffusion constant is adjusted for soil porosity, moisture content, and tortuosity (a calibration parameter which is a measure of irregularities in the gas-phase flowpath). Diffusion of CO_2 between soil layers in the liquid phase is negligible.

Alternatively, the model user has the option of setting the partial pressure of CO_2 or the liquid phase total inorganic C concentration, C_T, to a constant value in each layer.

4.8. NITRIFICATION

The rate of oxidation of ammonia to nitrate is represented by a Michaelis-Menten expression in which the rate of oxidation is linearly dependent upon the concentration of ammonia at low concentrations and is constant at high concentrations.

The decrease in the concentration of ammonia and the increase in the concentration of nitrate cause a decrease in the solution-phase alkalinity.

4.9. ANION ADSORPTION

The model simulates the adsorption of sulfate, phosphate, and organic acids as noncompetitive, reversible equilibrium processes. The adsorption of sulfate and phosphate is represented by linear isotherms which quantify the equilibrium distribution of the ions between the solution and sorbed phases ($C_{solution} = K \cdot C_{sorbed}$). Separate adsorption constants, K, can be used for each soil layer and each ion. The organic acid adsorption isotherm used ($C_{solution} = (K/\alpha) \cdot C_{sorbed}$, with $\alpha = \alpha_1 + 2\alpha_2 + 3\alpha_3$) is strongly dependent upon pH. The α_i's are the ionization constants for the model triprotic acid and are dependent only upon pH and temperature. This isotherm strongly favors adsorption as pH increases. Net adsorption of sulfate increases the alkalinity of the solution.

Although relatively simple adsorption isotherms are used, the model's conceptual formulation and numerical solution techniques readily accommodate other isotherms (e.g., Freundlich or Langmuir), including highly nonlinear expressions.

4.10. CATION EXCHANGE

Although mineral dissolution reactions (weathering) are the longterm producers of alkalinity and base cations, these reactions, in the absence of carbonate minerals, are slow. The ability of a soil to quickly neutralize acids and maintain relatively constant solution pH values comes from the large pool of base cations which exists on the cation exchange sites (this pool is large relative to the solution pool even in soils with low cation exchange capacities and low base saturation indices). These base cations can be quickly exchanged for H^+-ions.

The model simulates cation exchange as competitive, reversible, equilibrium reactions. Six cations are exchanged: Ca^{2+}, Mg^{2+}, K^+, Na^+, NH_4^+, and H^+. The exchange of cations of equal valence, for example, the exchange of Ca and Mg,

$$(Ca^{2+})_{ads} + (Mg^{2+})_{soln} \rightleftarrows (Ca^{2+})_{soln} + (Mg^{2+})_{ads}, \qquad (31)$$

is represented by mass action equilibrium equations.

The exchange of divalent cations with monovalent cations, for example, the exchange of Ca and H^+,

$$(Ca^{2+})_{ads} + 2(H^+)_{soln} \rightleftarrows (Ca^{2+})_{soln} + 2(H^+)_{ads}, \qquad (32)$$

is represented by empirical exchange equations, called Gapon equations, of the form,

$$\frac{\sqrt{\frac{[Ca^{2+}]_{soln}}{2}} [H^+]_{ads}}{[Ca^{2+}]_{ads} [H^+]_{soln}} = K_{Ca^{2+}, H^+} \tag{33}$$

where K_{Ca^{2+}, H^+} is the selectivity coefficient for H^+ relative to Ca^{2+}.

4.11. MINERAL WEATHERING

The ultimate production of alkalinity within a watershed originates from the weathering of primary minerals. These reactions consume H^+-ions and release base cations to solution:

Primary Mineral + xH$^+$ → Secondary Mineral (e.g., clays)

+ (x + y) Equivalents of Base Cations

+ y Equivalents of Strong Acid Anions

+ b H$_4$SiOH$_4$. (34)

The amount of H^+-ion consumed by weathering and the corresponding production of base cations is calculated in the model using specific mineral weathering stoichiometric coefficients (Newton and April, 1981) and rate expressions of the form shown in Equation (29).

4.12. ALUMINUM SYSTEM

The Al reactions are summarized in Figure 6. The protonation and complexation reactions with organic acid ligand, and SO_4^{2-}, and F^- are all fast solution-phase reactions which are represented mathematically by equilibrium expressions. The dissolution and precipitation of gibbsite (or any Al(OH)$_3$ solid) can be represented either by an equilibrium expression (solubility product), if the solution is in equilibrium with gibbsite (e.g., as occurs in the lower soil horizons), or by a rate expression if the solution is under or oversaturated with respect to gibbsite (e.g., in the organic horizon, the soil solution is typically undersaturated with respect to gibbsite solubility (Schofield, 1985). Under the rate-limited conditions, the following mass action rate expression is used to represent the change in total aqueous Al concentration:

$$R = k([Al_{T_{sat}}]_e - [Al_T]) \tag{35}$$

where

R = The rate of change in the concentration of total aqueous monomeric Al due to gibbsite dissolution precipitation;

k = Specific reaction rate constant;

$[Al_{T_{sat}}]_e$ = Total concentration of monomeric Al which would be in solution if the solution were in equilibrium with gibbsite (includes Al complexed with OH$^-$, SO_4^{2-}, F^-, and organic acid ligand); and

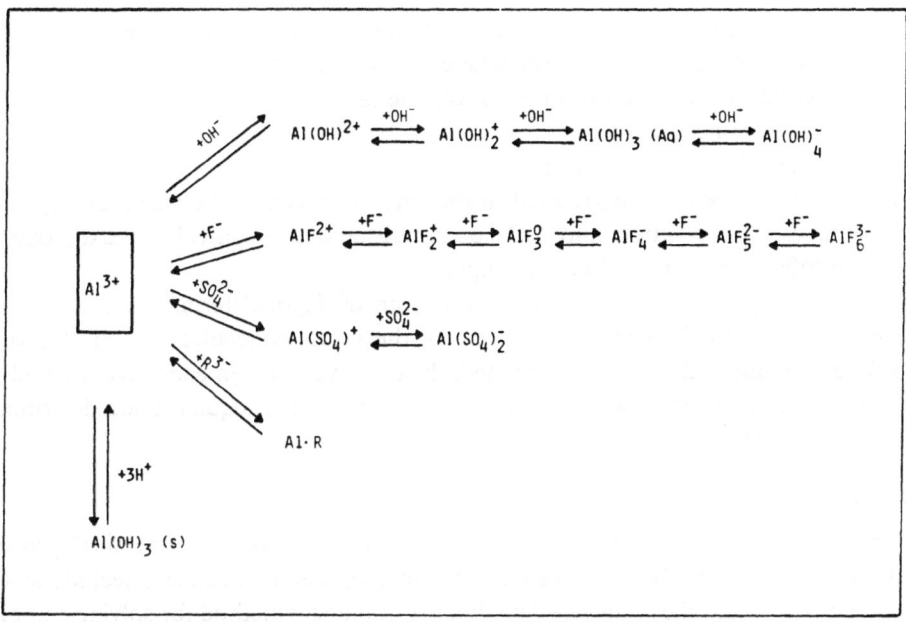

Fig. 6.   Aluminum dissolution and complexation with hydroxide, fluoride, sulfate, and organic acid ligand.

$[Al_T]$ = The actual monomeric Al concentration including complexes.

The concentration $[Al_{T_{sat}}]_e$ is a function of the thermodynamic constants for Al hydrolysis and complexation, and the concentrations of H^+, SO_4^{2-}, F^-, and organic acid.

4.13. SURFACE WATER PROCESSES

In streams, hydraulic residence times are short. Solute concentrations are calculated using the mass balance-based dilution method. Solution-phase equilibration is simulated as in the soils. Carbon dioxide is allowed to exchange with the atmosphere at a rate which is proportional to its difference from saturation. Heat is also exchanged with the atmosphere at a rate proportional to the difference between the stream's actual temperature and its equilibrium temperature.

To calculate the solute concentrations in the lake water, a mass balance equation is written for each horizontal layer of the lake, for each solute:

$$\frac{d(VC)}{dt} = \sum Q_i C_i + \sum Q_u C_u + \Sigma Q_0 C + AE \frac{dC}{dz} + \Sigma S$$

where
V = layer volume;
C = solute concentration in layer;
Q_i = volumetric inflow rate to layer i;
C_i = solute concentrations in the inflow;

Q_u = internal advective flow that enters the layer from adjacent layers;
C_u = solute concentration in layer where Q_u originates;
Q_0 = internal advective flow which leaves the layer;
A = layer interfacial area;
E = solute dispersion coefficient;
dC/dz = solute concentration gradient in the vertical direction (between layers); and
ΣS = fluxes of deposition and CO_2 exchange, and the equivalent fluxes due to nitrification and algal nutrient uptake.

A system of simultaneous linear equations with tridiagonal terms is developed for each solute. The equations are then solved for the solute concentrations in all layers, one solute at a time. After this is completed, the dissolved ions in each layer (including Al, organic acid, and carbonate species) are equilibrated. This equilibration determines the lake water pH.

4.14. SOLUTION TECHNIQUES

A variety of numerical solution techniques is used to solve the chemical process equations, including the Newton-Raphson technique, the delta-squared technique, the bisection method, and the regula-falsi method. A numerical method for solving complex expressions was developed by Lingfung Mok during the project. The method, based upon simultaneous interpolation and extrapolation, provides for rapid convergence with minimal computational effort. The method is used in the PHALKN subroutine which computes the solution-phase equilibria. This is the most frequently called subroutine and the method has cut model execution time by a factor of 3.

The solution technique, which allows the model to handle both equilibrium and nonlinear rate processes, results from the original model conceptualization of the natural system as a series of mixed reactors (CSTR's). The order of calculation over a time step is shown schematically in Figure 7. At the beginning of the time step, within each reactor (e.g., soil layer), there is an initial volume of liquid, V_i, and an initial mass of solute (e.g., Ca), M_i, and a solute concentration, C_i, which equals M_i/V_i.

The products of the slow reactions (e.g., weathering) are then added to the solution and a new unequilibrated mass of solute, M_i^* then exists in solution, where $M_i^* = M_i +$ (Rate of Production of Solute) $\cdot \Delta t$, where Δt is the length of the time step. Next, the external inputs are added to the solution. This increases the solution volume to V, where $V = V_i + Q_{in} \cdot \Delta t$, and the solute mass to M, where $M = M^* + (Q_{in} \cdot C_{in} + I) \cdot \Delta t$, with C_{in} being the solute concentration in the inflow and I being the input rate for solute not brought in with the water (e.g., dry deposition). The solution, solid, and gas phases are then equilibrated giving a new solution-phase solute mass, M_E, where $M_E = V \cdot C_E$. The outflows, Q_{out}, are then withdrawn at the equilibrium concentration, C_E, and at the end of the time-step, liquid volume, V_F, and total solute mass, M_F, are calculated: $V_F = V - Q_{out} \cdot \Delta t$ and $M_F = M_E - Q_{out} C_E \cdot \Delta t$. These calculations are performed for each of the solutes simulated for each time step, for each mixed volume (CSTR).

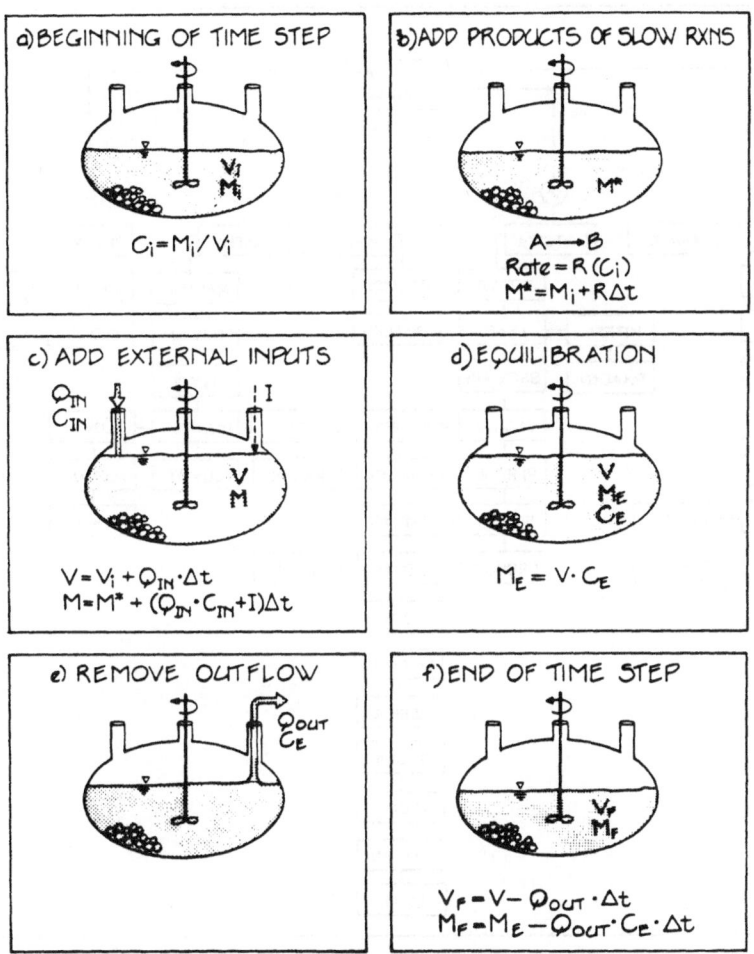

Fig. 7. Batch reactor analogy for CSTR over time step.

5. ILWAS model code

The model computer code is organized for modular execution. For example, the hydrologic module can be run separately from the chemical module; the snowpack can be simulated without including the soil and lake systems. Figure 8 shows the modular structure of the code. The code contains 51 subroutines and is about 12 700 lines long. The model was originally developed on a PRIME 550 computer with a FORTRAN IV compiler. A version of the code has been modified for IBM computer systems with FORTRAN 77 compilers. The conversion was verified and approved for release by the EPRI Software Distribution Center and Tetra Tech in April 1984. The model requires about 0.96 megabytes of core/virtual memory for execution. Execution times depend on

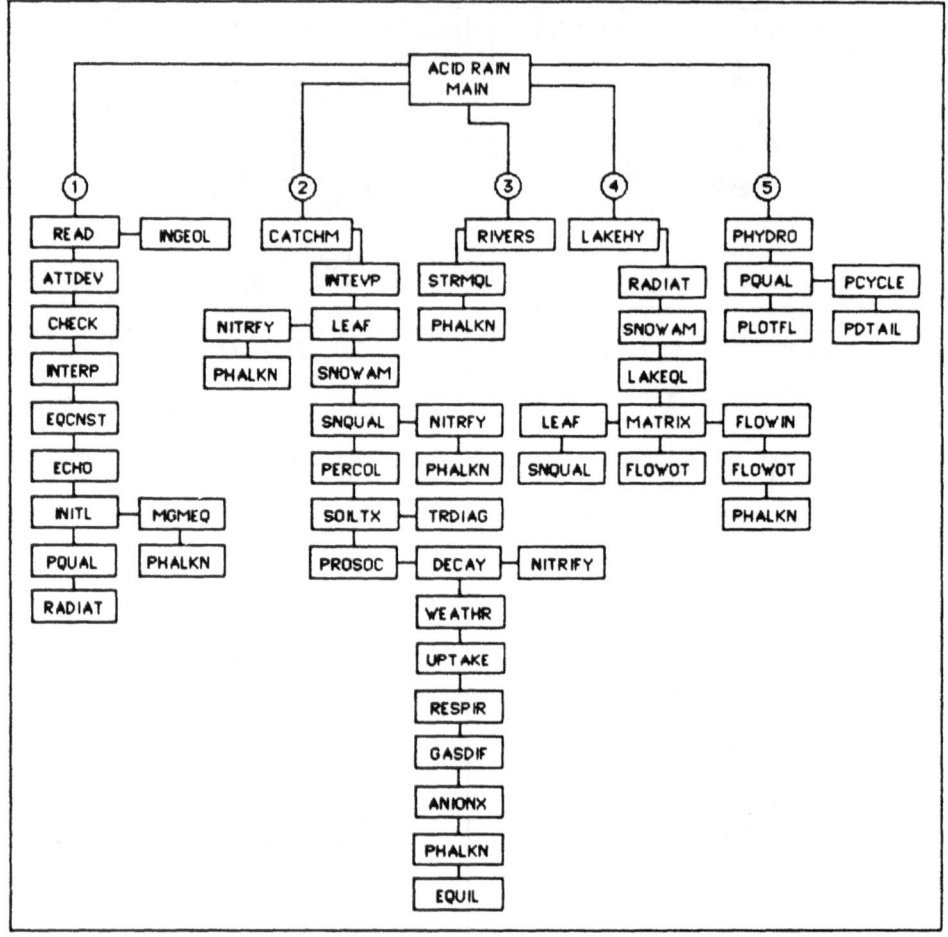

Fig. 8.   Relationship of ILWAS model subroutines.

the number of subcatchments used per hydrologic basin and the length of time simulated. Simulation of a nine-subcatchment lake-watershed for 1 yr takes less than 5 min of CPU time on an IBM 3081 computer and 138 min on a PRIME 550.

PART B. MODEL APPLICATION

6. Introduction

The ILWAS model can be applied to test scientific hypotheses and to answer management questions. In addition, the model can be used as a tool for checking the consistency of data collected by various investigators. For example, precipitation quantity and air temperature must be correct for the model to predict the amount of snowpack observed.

Hypotheses can be tested with the model by changing the calibration data, input data, or, in some cases, portions of the model itself. Management questions are addressed most commonly by changing model input (e.g., the acidity of precipitation) and observing model output (e.g., the acidity of lake water). Before the model can be so used, it must be calibrated. In the sections which follow, model calibration, hypothesis testing, and management questions will be addressed in detail.

7. Calibration

Application of the ILWAS model begins with calibration. Basin data are used to quantitatively characterize the system to be simulated. Initial conditions (e.g., lake stage, soil and surface water quality) are established as a simulation starting point. The model is then run using actual meteorological and air quality data as input. The model output, the quantity and quality of water at various points in the system, is made to coincide with observed values by adjustment of calibration parameters (e.g., evapotranspiration coefficients). The simulated results are typically compared to the observed data using graphical or statistical procedures. Simultaneous calibration against observed data for several points within a watershed is recommended if data are available. For example, throughfall quantity, snowpack depth, and flows at various points in any streams, should be calibrated together with lake discharge. Since the ILWAS model simulates many processes and contains several parameters, adjustment of these parameters must follow a logical procedure for calibration to proceed with minimal effort. The general rules for calibration are: (1) calibrate the system's hydrologic behavior before calibrating the chemical behavior; (2) calibrate in the same order as water flows through the basin; and (3) calibrate on an annual basis first, then seasonally, and finally calibrate to the instantaneous (daily) behavior.

In principle, some of the calibration parameters are measured values which cannot be changed arbitrarily. However, the measurements are commonly made at a limited number of locations (typically, one or two points), allowing some adjustment for calibration.

7.1. HYDROLOGIC CALIBRATION

The first step in the hydrologic calibration is the adjustment of the evapotranspiration parameters so that the predicted cumulative lake outflow matches the observed data. Observed lake outflow is checked against the observed rainfall to ensure that all major storm events cause either increases in lake outflow or at least retard recessions. The seasonal patterns in the predicted cumulative outflow are controlled by the length of the snow cover period and by seasonal variations in evapotranspiration. The snow formation temperature and the snowmelt rate coefficients, along with the sublimation rate, are adjusted to match observed snowpack depth and lake outflow during the snowmelt period. A seasonal calibration parameter is used to adjust the variation in potential evapotranspiration. Canopy interception storage, monthly leaf area indices, and root distribution determine the allocation of evapotranspiration to the different compartments of the system. Depending on the depth of the water table, the distribution of potential evapotranspiration will influence the actual amount of water lost (i.e., a large distribution factor for the upper soil layer together with a deep water table will give a low total evapotranspiration rate despite a high potential).

The peak flow and recession curve characteristics are influenced by the snowpack, soil field capacity, soil-specific yield, and soil hydraulic conductivity. The fine adjustment of the instantaneous lake discharge rate can be performed in conjunction with the chemical calibration. However, the ratio of base flow to the sum of shallow and surface flow should be reasonably estimated using flow separation techniques before proceeding with chemical calibration. Figure 9 shows the cumulative outflow calibration of Panther Basin and Figure 10 shows the instantaneous outflow calibration.

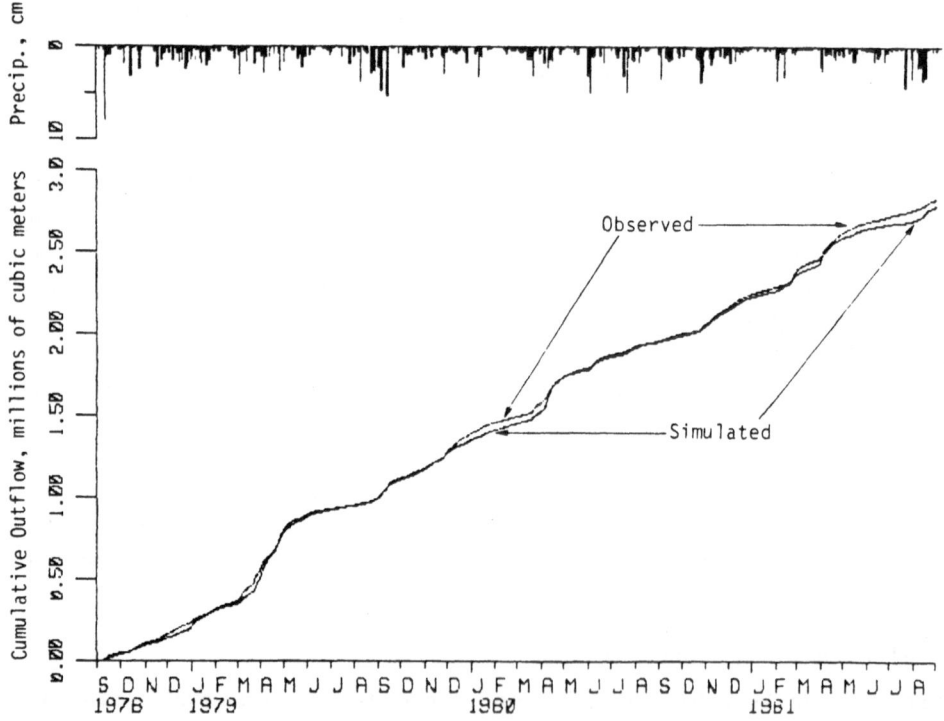

Fig. 9.   Simulated and observed cumulative outflow for Panther Lake.

7.2. CHEMICAL CALIBRATION

Lake chemical characteristics are primarily determined by the initial lake-water quality, in-lake processes, soil solution quality and the routing of water through the different soil layers. The quality of the soil solution is fairly constant with time because of equilibration of the aqueous phase with the solid phase by cation exchange and anion adsorption reactions. Hence, a reasonable calibration of lake outflow chemistry can be obtained

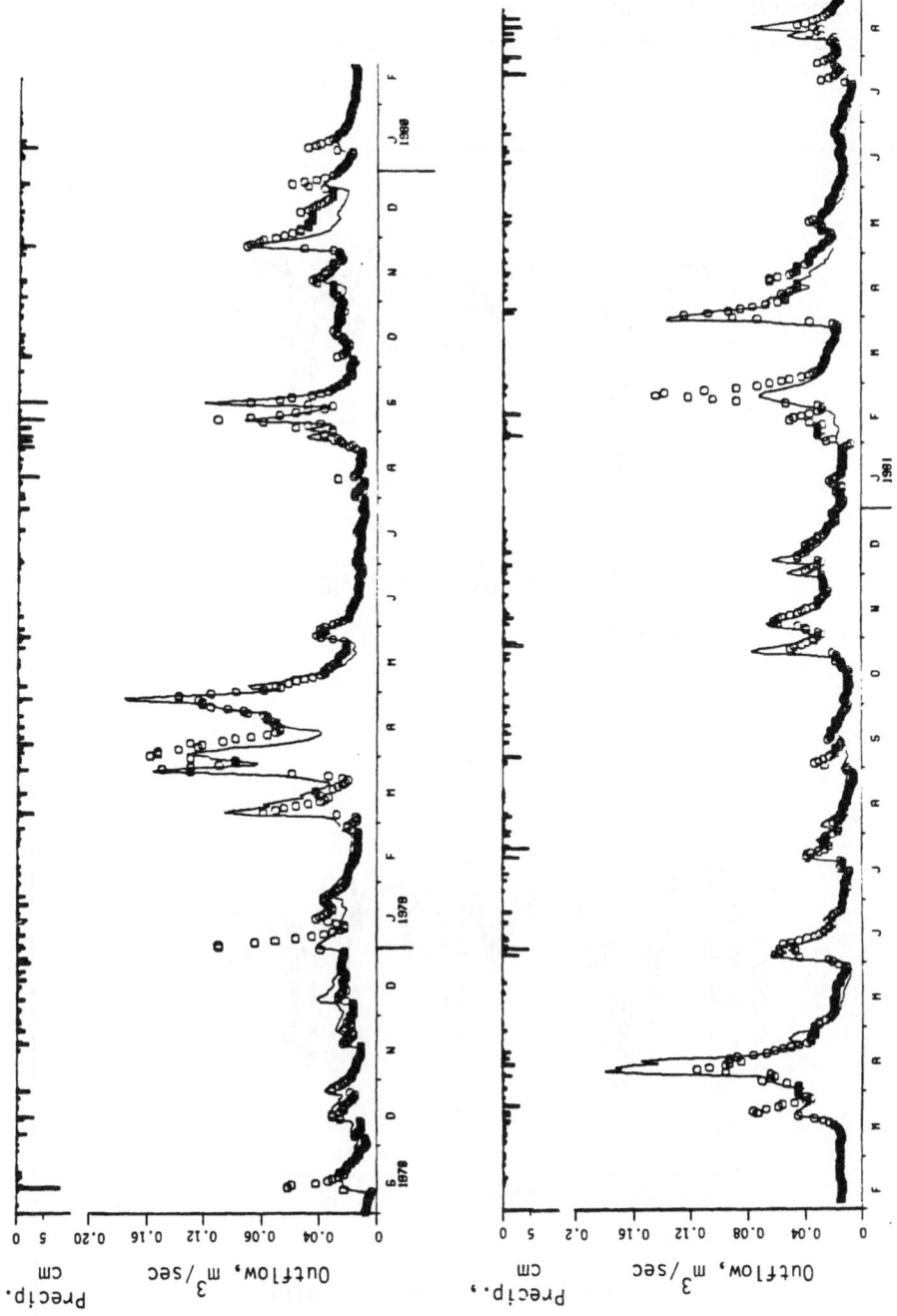

Fig. 10.   Simulated and observed instantaneous outflow for Panther Lake. Observed values are represented by small boxes, □.

by establishing the initial solute concentrations in the soil and lake waters, the soil cation exchange selectivity coefficients and anion adsorption coefficients, and the in-lake rate process coefficients. The initial solute levels in the soil solution must be such that a volumetric flow-weighted average of the soil solution concentrations, plus direct precipitation onto the lake surface, will give the average lake concentrations for species which

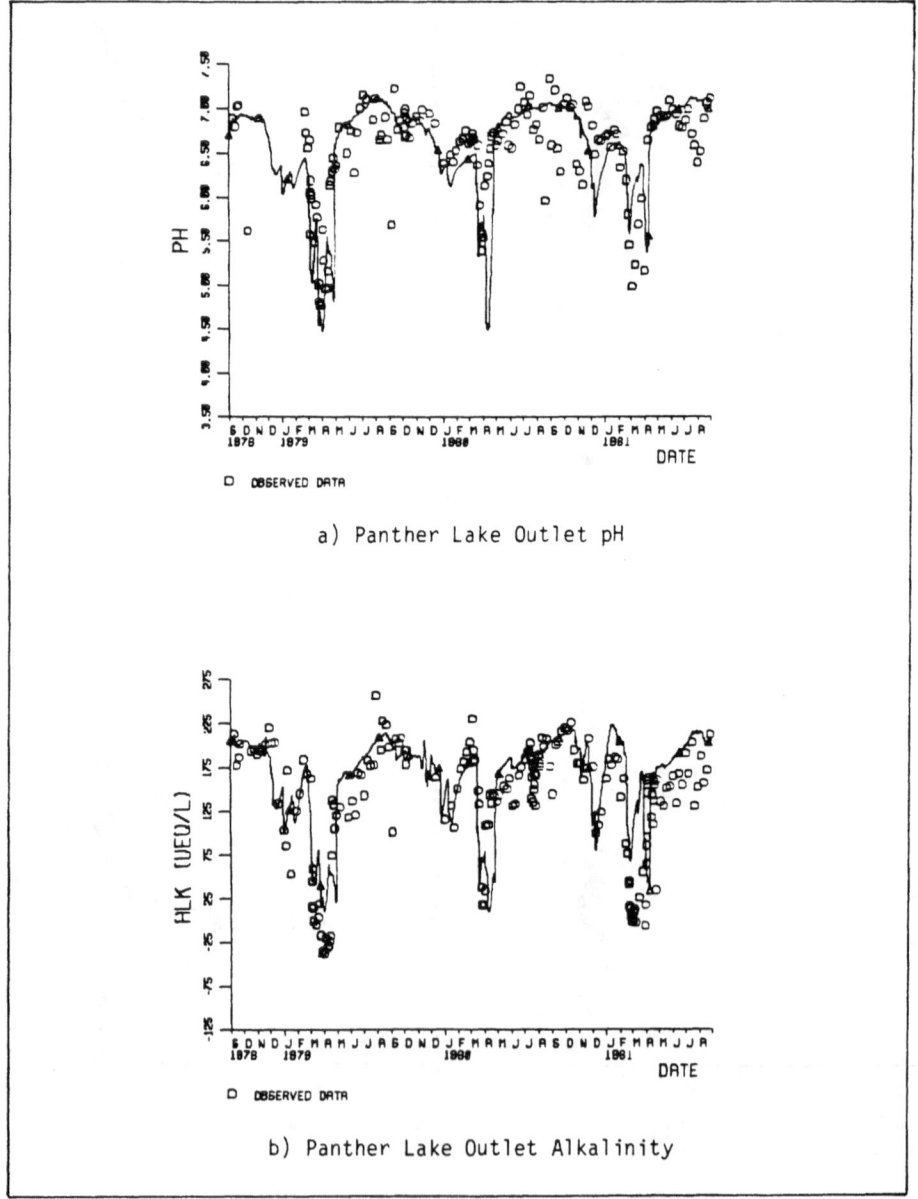

a) Panther Lake Outlet pH

b) Panther Lake Outlet Alkalinity

Fig. 11.   Simulated and observed outflow water quality for Panther Lake: (a) pH; (b) alkalinity.

do not undergo reaction in the lake (e.g., Cl^-). The volumetric flows used in the above averaging are the soil layer lateral flows obtained from the hydrologic calibration. The cation exchange selectivity coefficients and anion adsorption coefficients are adjusted so that the initial soil solution concentrations are at equilibrium with the adsorbed concentrations. Algal nutrient uptake and nitrification are the most important in-lake chemical processes currently represented by the model*. The soil nitrification rates, lake nitrification rates, and monthly algal production rates are adjusted to calibrate the ammonium and nitrate concentrations in the lake outflow.

Observed lysimeter and ground water chemical data can be used to check the hydrologic calibration. The concentrations of solutes which do not equilibrate with the solid phase may drift quickly away from their initial levels if the rate processes in the soil are not calibrated properly. The Al dissolution rate is adjusted to maintain a fairly constant level of Al in the applicable soil layers from year to year (Al concentration may fluctuate seasonally with the nitrate cycle). The nitrification rate is adjusted to prevent buildup of nitrate in deep soil layers. The soil tortuosity factor is adjusted to control the gas phase CO_2 diffusion rate and to maintain total inorganic carbon levels which are reasonably constant in the soil solution. Alternatively, the partial pressure of CO_2 or the total inorganic carbon concentration, C_T, can be set constant in the different soil layers. The observed silicic acid and Na^+ concentrations can be used in setting the mineral weathering rates.

If observed data are available, the throughfall chemistry can be calibrated by adjusting the gas and particulate deposition velocities, canopy nitrification rates, and foliar exudation rates. This also provides a means of quantifying total atmospheric deposition (wet and dry). The snowpack ion levels are calibrated by adjusting a solute leaching coefficient.

The dynamic behavior of the lake outflow water quality is influenced by the lake thermal profile, the fraction of solute retained in the lake ice, and the fraction of flow that comes in as surface runoff or ground water seepage. Figure 11 shows the calibrated and observed lake outflow pH and alkalinity for Panther Lake.

8. Hypothesis Testing

A major use of the ILWAS model is the testing of scientific hypotheses. In a sense, the model becomes a laboratory or field plot in which experiments can be conducted, including those which otherwise would be prohibitively costly or impossible. During the course of the study, three hypothesis-testing workshops were held at Tetra Tech's offices in Lafayette, California. Each five-day workshop was limited to 10 to 20 participants and included 7 to 8 of the principal investigators. On the first-day of the workshops, model theory and mathematical formulations were reviewed. On day two, hypothesis-testing procedures were discussed and the investigators developed specific hypotheses.

---

* In the RILWAS project (Goldstein *et al.*, 1985), additional in-lake processes, including redox reactions, have been added to the model.

STATEMENT OF HYPOTHESIS:_____ # 1
     Soil (till) depths control lake water acidity.*

EXPECTED RESPONSE: Lake water becomes more acidic as till thickness decreases and more alkaline as till thickness increases.

TESTING METHODS: Take Panther Basin, decrease the till depths. Take Woods Basin, increase the till depths. Simulate systems behavior for 5 years using the ILWAS Model.

RESULTS:

* Hypothesis originally formulated by R. Goldstein in early stage of ILWAS project.

Fig. 12. Example hypothesis statement form.

Included were clear statements of the hypotheses and expected results (Figure 12). From the 30 to 40 hypotheses formulated, the investigators select about 10 to 20 for actual testing. Over the next two days and nights, the selected hypotheses are tested using the ILWAS model and data base. The results are then reviewed on the fifth day.

A total of over 40 such hypotheses formulated by the principal investigators were tested using the ILWAS model. Included was a test of the major study hypothesis, that is, that the differences in the acidity of the surface waters of Woods and Panther Lakes are due to differing soil depths in the lakes' tributary watersheds (Goldstein *et al.*, 1984). On the average, Panther Lake basin has deep soil (average volume-weighted depth, 24 m), whereas Woods Lake basin has thin soil (average volume-weighted depth, 2 m)

(April and Newton, 1985). Basins with deep soils were expected to have more exchangeable base cations and weatherable minerals to neutralize acid deposition and therefore would be expected to have more alkaline surface waters. If the soil thickness of Panther basin was reduced to that of Woods, it was hypothesized that the lake water would become acidic. To complete the argument, it was hypothesized that Woods Lake would become alkaline if the soil thickness in its tributary watershed was increased to that of Panther.

Figure 13 presents the model results for the hypothesis described above. The average pH of Panther Lake was found to decrease to less than that of Woods Lake when the soil thickness in its tributary catchment was reduced to that of Woods by removing most of the underlying till.

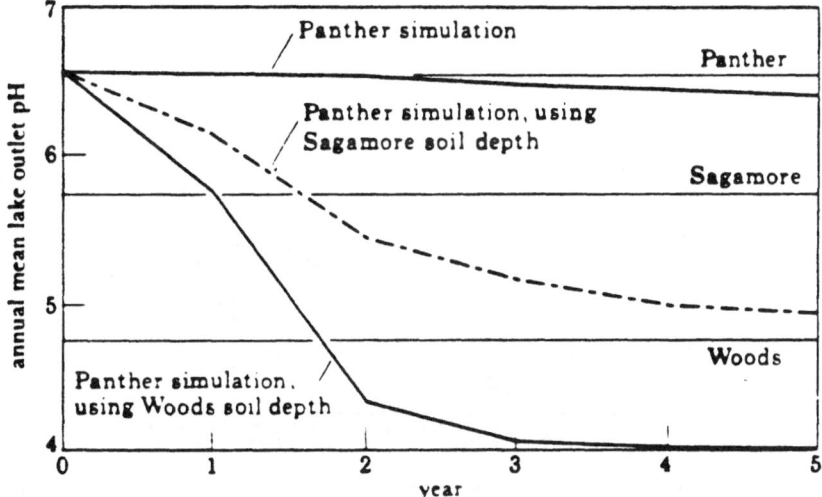

Fig. 13. Simulated annual mean lake outlet pH for Panther Lake using soil depths of Panther, Sagamore, and Woods watersheds. Horizontal lines are measured annual mean outlet pH values for the three lakes (Goldstein *et al.*, 1984).

Note that the Panther simulation using Woods basin soil thickness calculates a pH value lower than that of Woods Lake. This may be due in part to the difference in canopy distributions between the two basins. Panther basin has a higher percentage of coniferous trees than does Woods. Conifers have been shown to be more efficient collectors of dry deposition and dry deposition in the study area is acidic (Johannes *et al.*, 1985).

One may ask why soil depth should make a difference in lake water acidity. The answer is provided by the concept of flowpaths. Figure 3 presented the characteristic solution pH in soil layers found in the watersheds. The pH of the solution in the upper most horizon is noted to be lower than that of the incident precipitation. Further the pH values increase with soil depth. If the conditions were such that more water reaches the lake by way of the deeper soil horizons, the lake water would be more alkaline. If

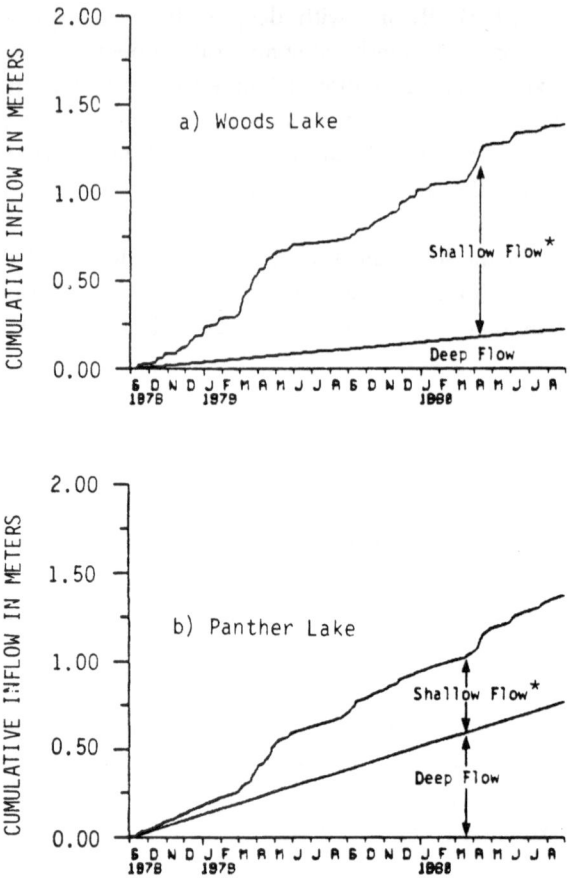

Fig. 14.   Cumulative lake inflows from Woods and Panther watersheds. Flows were predicted using the
ILWAS model. *Shallow flow includes surface runoff.

more water reaches the lake from lateral flow through the upper horizons, the lake water
would be more acidic.

Figures 14a and b show the amount of Woods and Panther lake inflow that is surface
runoff and shallow lateral flow through the upper organic soil horizons. These results
are derived from the model simulations. Based on these figures, Woods Lake appears
to receive shallow ground water inflow throughout the year and is therefore more acidic.
Panther Lake receives shallow ground water flow predominantly during the snowmelt
period. Its surface water, normally alkaline, becomes acidic during this period. It
appears that Panther Lake is more alkaline than Woods Lake because a larger
proportion of the precipitation falling onto its basin passes through its mineral soil
horizons.

A particularly interesting hypothesis was tested at the ILWAS Basin Input Workshop
in 1982. It was hypothesized that the springtime pH depression observed annually
during four previous years of sampling in Panther Lake was due to the surge of water

applied to the terrestrial system when the snowpack melts. It was hypothesized that if winter temperatures were higher, precipitation would be in the form of rain and would be applied more uniformly over time to the soil, the water would percolate deeper, and the pH depression in the lake would not occur.

This hypothesis was tested with the model by simulating the Panther Lake watershed for 1 yr using real input data except that the air temperature was never allowed to drop below 3 °C. This kept all precipitation in the form of rain and also prevented the formation of ice on the lake. The model results did not show the typical springtime pH depression in the lake surface waters; instead, the pH remained around 6.5 to 7 (See Figure 15). The model results at that point, though, were academic; no such conditions had been encountered in the field, and thus no confirmation was available.

By chance, the following winter provided a test of the model simulation. The winter was atypically warm. Little snowpack accumulated; only a slight buildup occurred late in the season. Continuing pH measurements at Panther Lake by Carl Schofield showed no significant springtime depression in pH!

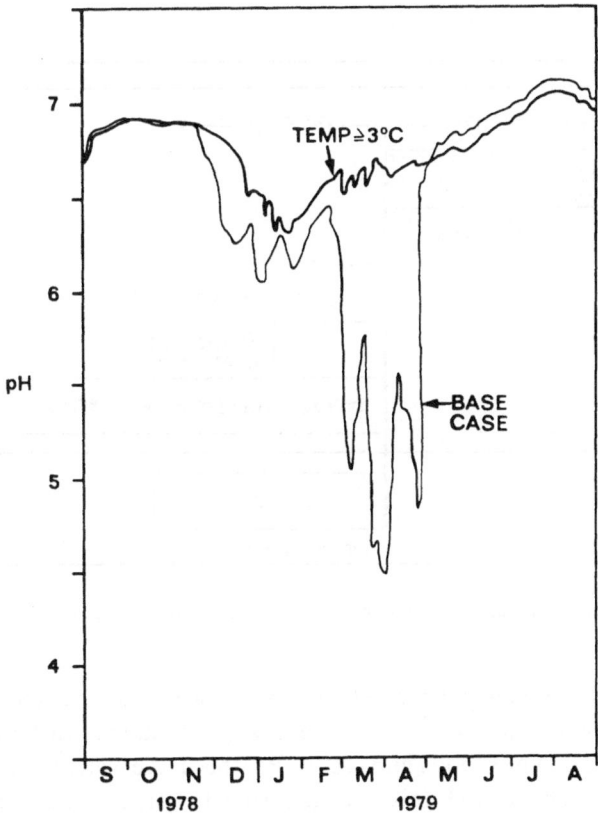

Fig. 15.   Simulated Panther Lake outlet pH for measured temperatures (base case) and for case where measured temperatures below 3 °C are set at 3 °C.

The model can also be used to compare the atmospheric inputs of acidity with the internal production of acids and bases within the basins. The results of such a comparison on an annual basis are shown in Figure 16 for Woods and Panther

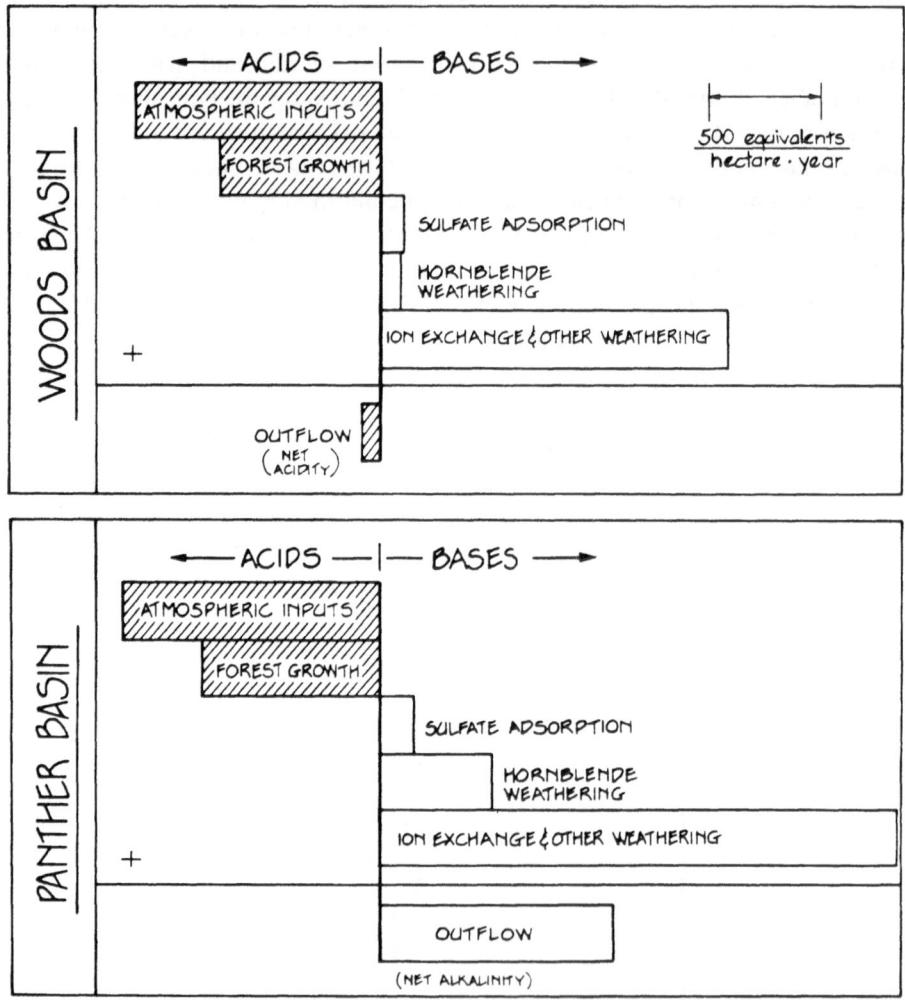

Fig. 16.   Annual acid-base budget for Woods and Panther lake-watersheds (Goldstein *et al.*, 1984).

lake-watersheds. The major difference between the two systems is the amount of base produced by 'other weathering and cation exchange reactions'. In Panther basin, this amount far exceeds that of Woods and causes Panther Lake water to have a net positive alkalinity and a much higher pH. It is also interesting to note that the internal production of acidity within the two basins is about the same and is about two-thirds as large as the atmospheric deposition of acidity.

9. Management Questions

In the ILWAS model, the major hydrologic and acid-base processes occurring in forested lake-watersheds have been mathematically abstracted. The inputs to the resulting coupled sets of equations can be easily changed to simulate system responses to perturbations. In this manner, the ILWAS model becomes a useful planning tool for answering 'what if?' questions. Information can be provided to support management decisions and the allocation of resources.

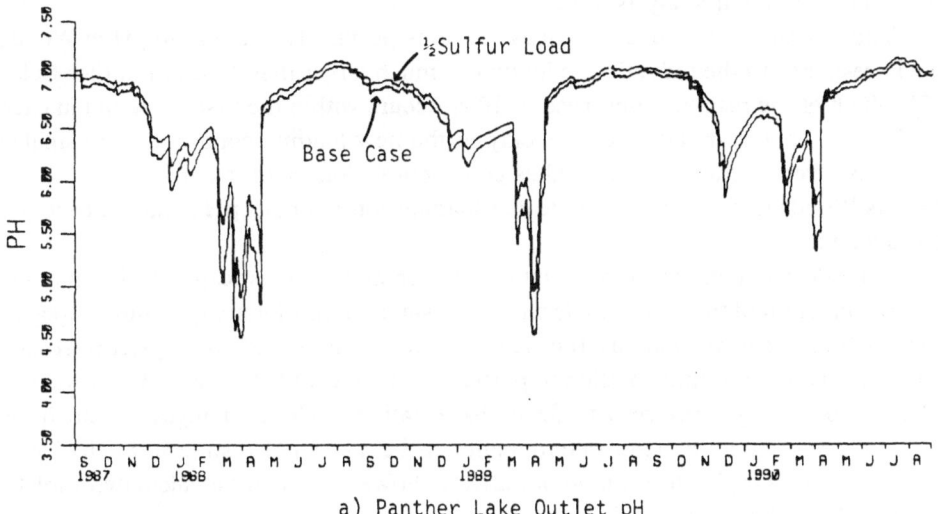

a) Panther Lake Outlet pH

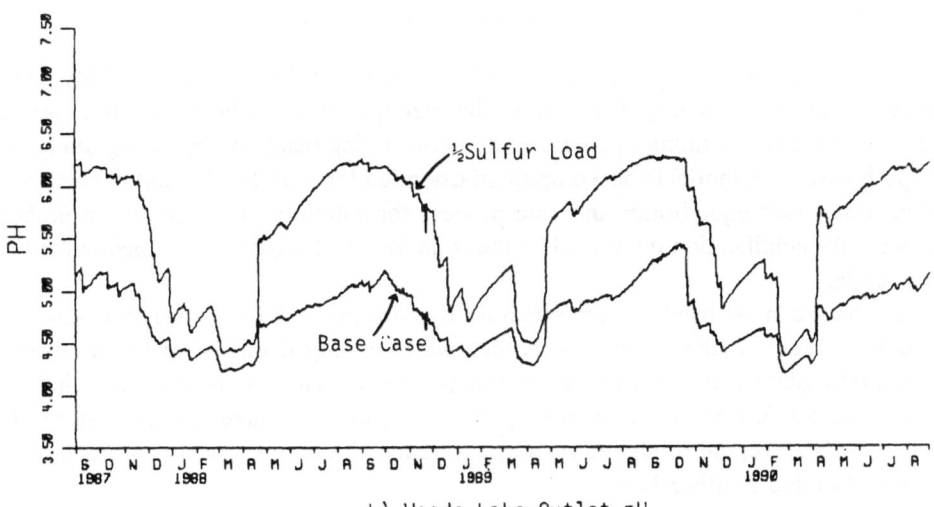

b) Woods Lake Outlet pH

Fig. 17. Response of Woods and Panther Lakes to a halving in the total atmospheric S deposition (wet and dry). Input data for 1978, 1979, and 1980 were run 'back to back' four times to create 12 years of simulation. The 3 yr shown above correspond to years 10, 11, and 12 of the simulation.

Among the most often asked management questions relative to lake acidification, are 'What will happen to lakes if the acidity of precipitation is increased or decreased?' and 'Over what time frames will any changes occur?'

These questions motivated the development of the ILWAS model. They can be addressed by changing the atmospheric deposition inputs to the model. Figure 17 shows the response of Woods and Panther lakes ten years after a halving of the total atmospheric S loading. The simulations shown were made using the input conditions for the period 1979 to 1981 repeated over and over to create a longer input record. These figures plus additional analysis indicate that:

(1) The responses to reduced loadings are lake-specific. The change in pH in Woods Lake in response to the reduction in loading is much larger than that in Panther Lake.

(2) Much of the response observed at 10 yr occurs within the first few years in each lake. Woods Lake responds more quickly to the new loading condition. The smaller change observed in Panther Lake takes considerably longer to occur.

(3) As Woods Lake responds to the new loading condition, its pH behavior becomes more dynamic.

These model simulations suggest that lake-watershed systems respond to changes in load. Examination of the model cycle charts for simulations of increased and decreased acidity loadings shows that, as the acid loading is increased, the systems release additional base cations into solution to partially offset the added acidity. As the loading is decreased, the systems release fewer base cations. These changes in the basin alkalinity supply rate are major determinants of the lake's pH response. Such system feedback is too complex to estimate intuitively; however, it can be calculated directly using the ILWAS model.

10. Summary

The ILWAS model was developed to predict changes in surface water acidity given changes in deposition acidity. It simulates the routing of water tributary to streams and lakes and the major alkalinity-producing and consuming reactions occurring along the flow pathways. The model is based upon an extended form of the alkalinity expression and includes both equilibrium and rate process formulations. It represents a unified theory of lake acidification quantified by the equations and algorithms embedded in the model code.

The model can be used to test scientific hypotheses and to answer management questions. Model results show that surface water quality depends on the flowpaths precipitation follows enroute to becoming surface water. The results also show that the lake-watersheds respond to changes in deposition acidity in a lake-specific manner. In acidic Woods Lake, the simulated response to reduced S deposition was larger than in the more alkaline Panther Lake.

References

Chen, C. W., Gherini, S. A., and Goldstein, R. A.: 1979, 'Modeling the Lake Acidification Process', in *Proceedings of Workshop on Ecological Effects and Acid Precipitation*, Central Electricity Research Laboratory, United Kingdom. Section 5, pp. 1–26.

Chen, C. W., Gherini, S. A., Hudson, R. J. M., and Dean, J. D.: 1983, *The Integrated Lake-Watershed Acidification Study:* Volume 1, Model Principles and Application Procedures. Electric Power Research Institute, Research Project 1109-5, EA-3221.

Gran, G.: 1952, *Analyst* **77**, 661.

Goldstein, R. A., Gherini, S. A., Chen, C. W., Mok, L., and Hudson, R. J. M.: 1984, *Phil. Trans. R. Soc. Lond.* **B305**, 409.

Johannes, A. H., Altwicker, E. R., and Clesceri, N. L.: 1985, *Water, Air, and Soil Pollut.* **26**, 339 (this issue).

Newton, R. and April, R.: 1981, personal communication.

Schofield, C.: 1986, *Water, Air, and Soil Pollut.* **26**, 403.

April, R. and Newton, R.: 1985, *Water, Air, and Soil Pollut.* **26**, 373 (this issue).

Stumm, W. and Morgan, J. J.: 1970, *Aquatic Chemistry – An Introduction Emphasizing Chemical Equilibria in Natural Waters*, John Wiley and Sons, Inc. pp. 1–583.

Valentini, Joy.: 1985, ILWAS Data Base, Tetra Tech, Inc., Lafayette, CA.

ANNOUNCEMENT

THIRD INTERNATIONAL SYMPOSIUM ON ENVIRONMENTAL MANAGEMENT FOR DEVELOPING COUNTRIES

The Third International Symposium on Environmental Management for Developing Countries will take place in Istanbul, Turkey, between August 6–12, 1986.

The following subjects will be covered:
- Appropriate technology for liquid, solid and gaseous waste treatment and disposal in touristic areas
- Effects of industrial activities on tourism
- Effect of environmental conditions on tourism
- Effects of touristic activities on environment
- Rehabilitation of touristic areas destroyed by industrial and touristic activities
- Socio-economic aspects of tourism, industry and environment interaction
- Water supply and waste management thoughout history (water and wastewater collection, treatment and disposal methods in old communities)
- Sanitation facilities in yachts, ships, etc. and harbours and marine
- Sampling, characterization and monitoring of hazardous wastes
- Environmental and health impacts of hazardous wastes
- Methods for minimizing hazardous waste generation
- Legislation and regulations related to hazardous waste treatment and disposal of hazardous and difficult to be handled solid, liquid and gaseous wastes
- Case studies.

Papers describing techniques appropriate for developing countries will have a priority.
Engineers and scientists of any nationality are welcome to submit papers or participate in the symposium. Authors are invited to submit a copy of a maximum 500-word abstract by December 15, 1985. For further information please contact:

ENVITEK
Environmental Technology, Research and Development Center

Bahariye Cad. 56
Kadıköy-Istanbul, Turkey
Telephone: (90–1) 3364795
Telex: 29505 ktk-tr
             attn. Envitek 105

Cable: ENVITEK-Kadıköy-Istanbul/TURKEY

ANNOUNCEMENT

Integrated Lake-Watershed Acidification

Please note that a hardbound edition of this special issue *Water, Air, and Soil Pollution,* Vol. 26, No. 4, (December 1985) is available from the publishers.

ISBN 90-277-2162-9   Prices: Dfl. 75,– $29.50   £20.95

AUTHOR INDEX
Volume 26

SUBJECT INDEX

Volume 26

TABLE OF CONTENTS

Volume 26 – 1985

TABLE OF CONTENTS

TABLE OF CONTENTS

TABLE OF CONTENTS

TABLE OF CONTENTS